Baboon Metaphysics

Baboon Metaphysics

The Evolution of a Social Mind

DOROTHY L. CHENEY
AND
ROBERT M. SEYFARTH

The University of Chicago Press   Chicago and London

DOROTHY L. CHENEY is professor of biology and ROBERT M. SEY-
FARTH is professor of psychology at the University of Pennsylvania.
They are the authors of *How Monkeys See the World,* also published by
the University of Chicago Press.

The University of Chicago Press, Chicago 60637
The University of Chicago Press, Ltd., London
© 2007 by The University of Chicago
All rights reserved. Published 2007
Printed in the United States of America

16 15 14 13 12 11 10 09 08 07   1 2 3 4 5
ISBN-13: 978-0-226-10243-6 (cloth)
ISBN-10: 0-226-10243-2 (cloth)

Library of Congress Cataloging-in-Publication Data
Cheney, Dorothy L.
   Baboon metaphysics : the evolution of a social mind / Dorothy L.
Cheney and Robert M. Seyfarth.
      p.   cm.
   Includes bibliographical references and index.
   ISBN-13: 978-0-226-10243-6 (cloth : alk. paper)
   ISBN-10: 0-226-10243-2 (cloth : alk. paper)
   1. Baboons—Behavior.   2. Social behavior in animals.
3. Cognition in animals.   I. Seyfarth, Robert M.   II. Title.
   QL737.P93C43   2007
   599.8'65—dc22

                                        2006102646

Contents

Acknowledgments

Our research would not have been possible without the help and encouragement of Bill Hamilton, who in 1992 generously allowed us to assume charge of his camp and C troop, the baboon group he and his colleagues had studied since the late 1970s. Bill died in 2006, after a distinguished career as Professor of Zoology at the University of California, Davis. He was a superb naturalist with a talented appreciation for animals and their environment.

We owe a huge debt of gratitude to the Office of the President, the Department of Wildlife and National Parks, and more recently, the Ministry of Environment, Wildlife and Tourism of the Republic of Botswana for granting us permission to work and live in the Moremi Game Reserve, one of the most spectacular places on earth. We especially thank Keith Lindsay, Doug Crowe, Joe Matlhare, Jan Broekhuis, Daniel Mugogoho, Rapalang Mojaphoko, and Cyril Taolo in the Department of Wildlife and National Parks for their encouragement and sponsorship of our research, and the staff in the Ministry of Environment, Wildlife and Tourism, particularly Mrs. Maswego Maswabi, for their help and efficiency.

We also thank the various organizations that have supported our work over the past fifteen years: the U.S. National Science Foundation (NSF) and National Institute of Mental Health (NIMH), the National Geographic Society, the Leakey Foundation, and the University of Pennsylvania. We are especially indebted to Howard Kurtzman, our program director at NIMH, who intervened to preserve a large portion of our final grant.

We doubt if anyone has ever abused the pronoun "we" as promiscuously as we have in this book. In fact, we would not have been able to conduct our research, much less contemplate this book, had it not been for the collective efforts of the many friends who joined us on this project, gathering much of the data and conducting many of the experiments: Joan Silk, Robert Boyd, Ryne Palombit, Drew Rendall, Karen Rendall, Sarah Johnson, John Bock, Julia Fischer, Kurt Hammerschmidt, Marcus Metz, Dawn Kitchen, Jim Nicholson, Jacinta Beehner, Thore Bergman, Anne Engh, Rebecca Hoffmeier, Cathy Crockford, Roman Wittig, and Eva Wikburg. Space permits us to list only a few of their many and varied talents: Joan's ability to use chocolate as an inducement to lure small children into forced marches through hot sand; Robert's epic tales during long distance runs; Sam Boyd's willingness to offer himself up for sacrifice; Kristin Palombit's cello concerts for baboons; Julia's ability to send people into the sky when annoyed; Dawn's sponsorship of the ladies' letaka weaving cooperative's tea; Jim's creativity with pellet guns, duct tape, and maglites; Jacinta's aptitude for speaking only in bursts of dialogue from *Pride and Prejudice* for days on end and her design of the *Baboons versus Researchers* board game; Thore's tenure as a role model for Maun's wayward youth; Anne's creation and bottling of Queen of Mean chardonnay; Cathy's aplomb during her stormy relationship with Nat; and Roman's documentation of tool use in baboons. We also thank Selo for maintaining group order, Martha for her leopard-mobbing skills, Elvis for his family values, and Comet for returning the fishing rod.

Special thanks go to our research assistants, Mongabe Kgosiekae, Mpitsang Mokopi, Mokopi Mokopi, and Mange (Alec) Mokopi. Mongabe worked for Bill Hamilton for many years, helped us at the start of our project, and then retired, turning over his duties to three brothers: Mpitsang, a tireless worker in camp, Mokopi, an extraordinarily gifted naturalist who died in 2005, and Alec, an equally talented and reliable field assistant. Mokopi's and Alec's skills in tracking animals saved us on many occasions from stumbling onto lions, buffalo, and elephants, and their uncanny ability to recognize each baboon—often by voice alone—helped us to keep track of who was doing what to whom even as the protagonists raced wildly through trees and shrubs. Mokopi and Alec also provided insights into the parallel universe inhabited by hippos, Rombe, Chris, Other Guy, Mr. Nature, and the staff at the Maun hospital, and they collected fecal samples with equanimity.

Our research has relied enormously on numerous people in Maun and on the managers and staff of Delta and Xaxaba (Eagle Island)

Camp, the two tourist lodges downriver from us. In each case we fear that our parasitic dependence has tested the limits of friendship. In Maun we thank the staff and pilots at Mack Air, especially Stuart and Lara Mackay, for their logistical support, flexibility, and understanding over the years. We are also grateful to Dan Rawson for supply runs in Big Blue and for resuscitating our recalcitrant boat engines, Allie van Niekerk for providing a friendly and helpful voice on the other end of the radio, and Willy van Niekerk for his truck runs and anecdotes. At Delta Camp we thank Binky Newman and Robert Avelino Sekeseke for their help and hospitality. At Xaxaba (Eagle Island) Camp we thank the many managers and staff who not only helped us fight fires, cut river channels, and transport propane and petrol, but also entertained us with booze cruises, potjie bakeoffs, and tales of exasperating tourists bent on fulfilling their national stereotypes. We are especially indebted to Lecco Masoko, Johan and Leanne Pretorius, Guy Dudley and Lisle Bester-Dudley, Mike and Sally Oughtibridge, and Werner Smith and Chantelle Shaw.

Special thanks go to Ian Clark, who has almost single-handedly helped us to keep the camp supplied, and (more importantly) generously shared his explorations of the Okavango Delta with us. Ian not only introduced us to Tchau, the Xo Flats, and the three pleasure spots known as Pot, Weed, and Gilligan's Islands, but also demonstrated how to balance a boat on the top of a hippo's back and then slide it deftly back into the water.

Supply runs to Maun would have been dull indeed without the hospitality and great friendship of Tim and Bryony Longden, who, together with their children Maxie, Pia, and Blyth, provided us with a second home. Tim and Bryony introduced us to many eccentric people of uncertain provenance, organized the first and last annual Baboon Night at the Duck Inn, and inspired us with their indomitable optimism and love for rain, the flood, the delta's plants and animals, and the seven-club bid. They also taught us that minor prangs with microlites, cars, recalcitrant horses, and disgruntled ostriches only enhance one's appreciation of life.

Over the years, we and the other denizens of Baboon Camp have also enjoyed the Maun friendship and hospitality of Colin McAllister and Jane Rawle, Ellie van Keken, and Ruth Stewart and Sande Greer. We are especially grateful to Tico McNutt and Lesley Boggs. At the start of our study in 1992 Tico helped us rebuild Baboon Camp, and in the years since then Tico and Lesley have shared with us their research on wild dogs and convinced us of the necessity of playing petanque with boules. They also generously housed Keena and Lucy in their icebox for two months.

In Namibia, Conrad Brain and Ginger Manley provided us with hospitality and guidance in our quixotic quest for a goat-herding baboon and were tactful enough to pretend to take it seriously.

Our book has benefited tremendously from the detailed and critical comments of Joan Silk and Julia Fischer, who meticulously read an entire first draft, gently reaming out and disciplining our prose and logic. Others who read every chapter and provided insightful comments were Jacinta Beehner, Thore Bergman, Delphine Dahan, Sandy Harcourt, Dawn Kitchen, Alex Martin, Bob Martin, Keena Seyfarth, Lucy Seyfarth, Klaus Zuberbuhler, and several anonymous reviewers. We are also grateful to Christie Henry, Michael Koplow, Ryan Li, and many others at the University of Chicago Press for their help, advice, and patience.

Finally, we would like to thank Keena and Lucy, who have accompanied us along every step of this bizarre journey. From early years dedicated to stuffed animal operettas, francolin trapping, millipede races, pet leeches, and parlor games to later ones taken up by poop patrols, focal animal samples, hormone extraction, and ever more elaborate parlor games, they have helped us in ways too numerous to describe. With grace, wit, and aplomb they have mended plumbing destroyed by elephants, extracted cars from deep sand, changed countless tires, made solitary runs through hippo gauntlets, been cheerfully treed by a variety of different animal species, and endured innumerable arcane and inane conversations and visitors. They love Baboon Camp as much as we do; no one could ask for better friends.

ONE

The Evolution of Mind

Origin of man now proved.—Metaphysic must flourish.—He who understands baboon would do more towards metaphysics than Locke.
CHARLES DARWIN, 1838: *NOTEBOOK M*

What goes through a baboon's mind when she contemplates the 80 or so other individuals that make up her group? Does she understand their social relations? Does she search for rules that would allow her to classify them more easily? Does she impute motives and beliefs to them in order to better predict their behavior? Does she impute motives and beliefs to *herself* when planning a course of action? In what ways are her thoughts and behavior like ours, and in what ways—other than the obvious lack of language and tools—are they different? These are questions that also vexed Charles Darwin.

We have taken our title from one of Darwin's most memorable remarks. He wrote it on August 16, 1838, almost two years after returning from his voyage on the *Beagle* and 21 years before the publication of *The Origin of Species*. It was a time of vigorous intellectual activity, when Darwin read voraciously on many subjects, both within and beyond the sciences, and met and talked with many different people, from family friends to prominent literary and political figures (Hodge 2003). Despite this active intellectual life, however, it seems unlikely that he or anyone else had ever combined the words "baboon" and "metaphysics" in the same sentence. What was Darwin thinking?

Mind and behavior in Darwin's time

The Cambridge English Dictionary defines metaphysics as "the part of philosophy that is about understanding existence and knowledge." Writing in the *Westminster Review* in 1840, John Stuart Mill offered a summary of views on the origin of knowledge that were being discussed by Darwin and his contemporaries. "Every consistent scheme of philosophy requires, as its starting point, a theory representing the sources of human knowledge, and the objects which the human faculties are capable of [understanding]. The prevailing theory in the eighteenth century ... was that proclaimed by Locke, and attributed to Aristotle—that all our knowledge consists of generalizations from experience" (Mill 1840). According to this theory, Mill continued, we know "nothing, except the facts which present themselves to our senses, and such other facts as may, by analogy, be inferred from these. There is no knowledge *a priori*; no truths cognizable by the mind's inward light and grounded on intuitive evidence." Locke believed that the mind acts simply to associate events that have been joined together through proximity and repetition. From these associations it generates behavior. Anything we think or do can ultimately be traced to our experience.

Mill continued: "From this doctrine Coleridge with ... Kant ... strongly dissents. ... He distinguishes in the human intellect two faculties ... Understanding and Reason. The former faculty judges of phenomena, or the appearance of things, and forms generalizations from these: to the latter it belongs, by direct intuition, to perceive things, and recognize truths, not cognizable by our senses." In Kant's scheme, these perceptions exist a priori but are not completely innate because they require experience for their expression. For Kant, the mind was not a blank slate on which any sort of experience can write any kind of instructions. It is, instead, biased in the way it responds to features of the world—actively organizing experiences and generating behavior on the basis of preexisting schemes. To understand our thoughts, beliefs, and behavior, therefore, we must consider not only our own individual experiences but also the preexisting nature of the mind itself.

Empiricism and rationalism were hotly debated at the time. Mill reported that "between the partisans of these two opposite doctrines there reigns a *bellum internecinum* [in which] even sober men on both sides take no charitable view of each others' opinions." Darwin followed the debate, but with a more open mind and a much more zoological perspective than many of his contemporaries. While others debated the nature of the human mind, he also puzzled over the minds of bees, dogs, and baboons.

Darwin's interest in metaphysics was motivated by more than just idle curiosity—it was also fueled by excitement and personal ambition. By the late 1830s and 1840s, the theory of evolution by natural selection was beginning to take shape in his mind, and his notebooks are filled with many speculations about how his work might shed an entirely new light on the study of human knowledge.

Darwin had observed that every animal species engages in repeated, "habitual" behavior. Birds build nests, squirrels hoard seeds, and dogs raise the fur on their back when they feel threatened. He believed that these behaviors recurred because they were beneficial to the individuals involved and that, over generations, habitual behavior became "instinctive," or innate. Under the right conditions, instinctive behavior would appear automatically, even if the animal had never before had the appropriate experience. When they act by instinct, then, animals are not behaving according to Lockean reason, carefully weighing the information acquired from experience. Instead, they are governed by "hereditary tendencies" acquired over generations (Darwin 1838a; for Darwin's views on habitual and instinctive behavior, see his other notebooks in P.H. Barrett et al. 1987).

This is not to say that Darwin believed animals were slaves to their instincts, wholly devoid of learning or reason. Some of his contemporaries did hold such views, and used them to draw a sharp distinction between humans and other animals. The naturalist Edward Blyth (1837), for example, wrote that "whereas the human race is compelled to derive the whole of its information through the medium of its senses, the brute is, on the contrary, supplied with an innate knowledge of whatever properties belong to all the natural objects around." Darwin disagreed—both with the conclusion that animals' thoughts and behavior are entirely based on instinct and with the view that human thought and behavior are governed entirely by reason. "[It is] hard to say what is instinct in animals & what [is] reason, in precisely the same way [it is] not possible to say what [is] habitual in men and what reasonable. ... as man has hereditary tendencies, therefore man's mind is not so different from that of brutes" (Darwin 1838a). Like many of his contemporaries, Darwin was searching for an explanation of mind and behavior that would combine innate, inherited tendencies (a bit of rationalism from Kant) with reasoning based on experience (a bit of empiricism from Locke) (Richards 1987). In this as in so much else, Darwin was a man ahead of his time.

3

Darwin also realized that, whatever the exact balance between innate behavior and reason in any particular instance, his theory of evolution had important implications for the study of metaphysics. After all, thoughts and instincts came from the mind, and the mind could be studied like any other biological trait. It was different in different species, reflecting the particular adaptations of each, and it could change gradually over time, being transmitted from one generation to the next. In his notebook M (M for metaphysics), Darwin wrote: "We can thus trace causation of thought ... [it] obeys [the] same laws as other parts of structure" (Darwin 1838b).

With growing excitement, Darwin began to see that his theory might allow him to reconstruct the evolution of the human mind and thereby resolve the great debate between rationalism and empiricism. The modern human mind must acquire information, organize it, and generate behavior in ways that have been shaped by our evolutionary past. Our metaphysics must be the product of evolution. And just as the key to reconstructing the evolution of a whale's fin or a bird's beak comes from comparative research on similar traits in closely related species, the key to reconstructing the evolution of the human mind must come from comparative research on the minds of our closest animal relatives. "He who understands baboon would do more towards metaphysics than Locke."

Twentieth-century views: behaviorists and their critics

In the first half of the 20th century, research on the mind and behavior was dominated by modern-day empiricists like E. L. Thorndike, J. B. Watson, and B. F. Skinner, who together developed the doctrine of behaviorism. Like Locke, they believed that organisms come into the world with little a priori knowledge: behavior is the product entirely of experience. As an animal moves through its world, it encounters stimuli and responds to them. If its response is followed by something pleasant, like food, the response will be repeated whenever the animal encounters the same stimulus again. In this way, the animal quickly develops an array of behaviors that are well suited to its needs.

As the intellectual descendants of Locke, behaviorists believed that the mind is concerned primarily with the formation of associations: mechanical principles of attachment that develop as a result of experience. They saw the mind not as an active "thinking" organ, predisposed to organize incoming stimuli in certain ways, but instead as a rather passive arena in which stimuli from the environment are combined according to simple rules, thereby producing behavior. The behaviorists concluded that a few simple but powerful laws, like Pavlov's Law of Association and Thorndike's Law of Effect, could account for all behavior, in every species and every circumstance. They believed in the principle of *equipotentiality*. As Skinner famously remarked, "Pigeon, rat, monkey, which is which? It doesn't matter ... once you have allowed for differences in the ways they make contact with the environment, what remains of their behavior shows astonishingly similar properties" (Skinner 1956:230–231).

The behaviorists saw little point in considering mental activities like thoughts, feelings, goals, or consciousness, for reasons that were both methodological and deeply philosophical. On the practical side, mental states like thoughts or emotions are private. They cannot be observed or measured, nor can one predict how they might be changed by experience. Under these circumstances, the mental activities of animals can hardly play a role in any scientific discipline. Even in humans, where introspection prompted some behaviorists to admit—grudgingly—that mental states might exist, the exact nature of these states are unknowable because they can never be verified by more than one person. Once again, this makes mental states unsuitable for scientific study. Some behaviorists went even further. In his 1974 book *About Behaviorism,* Skinner distinguished between "methodological behaviorists" who accepted the existence of mental states but avoided them because they could not be studied scientifically, and "radical behaviorists" like himself, who believed that "so-called mental activities" were an illusion—an "explanatory fiction." For Skinner, thoughts, feelings, goals, and intentions played no role in the study of behavior because they did not, in fact, exist.

Although behaviorism dominated 20th-century psychology, it was not without its critics. Perhaps the best way to understand them is to consider some classic observations and experiments that challenged the behaviorists' worldview.

Song sparrows (*Melospiza melodia*) and swamp sparrows (*Melospiza georgiana*) are two closely related North American birds with very different songs. Males in both species learn their songs as fledglings, by listening to the songs of other males. But this does not mean that the mind of a nestling sparrow is a blank slate, ready to learn virtually anything that is written upon it by experience. In fact, as classic research by Peter Marler and his colleagues has shown, quite the opposite is true. If a nestling male song sparrow and a nestling male swamp sparrow are raised side-by-side in a laboratory where they hear tape-recordings of both species' songs, each bird will grow up to sing only the song of its own species (Marler and Peters 1989).

The constraints that channel singing in one direction rather than another cannot be explained by differences in experience, because each bird has heard both songs. Nor can the results be due to differences in singing ability, because both species are perfectly capable of producing each other's notes. Instead, differences in song learning must be the result of differences in the birds' brains: something in the brain of a nestling sparrow prompts it to learn its own species' song rather than another's. The brains of different species are therefore not alike. And the mind of a nestling sparrow does not come into the world a tabula rasa—it arrives, instead, with genetically determined, inborn biases that actively organize how it perceives the world, giving much greater weight to some stimuli than to others. One can persuade a song sparrow to sing swamp sparrow notes, but only by embedding these notes into a song sparrow's song (Marler and Peters 1988). It is almost impossible to persuade a swamp sparrow to sing any notes other than its own (Marler and Peters 1989). Philosophically speaking, sparrows are Kantian rationalists, actively organizing their behavior on the basis of innate, pre-existing schemes.

In much the same way, human infants have their own sensory and cognitive biases. From the first days of life, they attend more readily to faces than to other visual stimuli and more readily to speech than to other auditory stimuli. This latter bias can apparently be traced to a preference for the intonation contours in spoken language: two-day-old babies show distinctive cerebral blood flow when they hear a normal sentence but not when the same sentence is played backward (Dehaene-Lambertz et al. 2002; Peña et al. 2003). Humans and sparrows are not alone in preferring their own species' sounds: when a rhesus macaque monkey (*Macaca mulatta*) hears a call given by a member of its own species, its brain exhibits activity that is markedly different from that shown in response to other sounds. Indeed, rhesus calls activate in the rhesus brain the same areas activated by human speech in the human brain (Gil da Costa et al. 2004).

Some of the most striking evidence for an innate predisposition to learn one's own species' communication comes from children who are born blind or deaf. Although they cannot see the objects in the world to which spoken words refer, blind children develop language at roughly the same age and in the same manner as children who can see (Landau and Gleitman 1985). Data from children born deaf are even more striking. Lila Gleitman, Susan Goldin-Meadow, and their colleagues studied several deaf children born to hearing parents who did not themselves know ASL, the American Sign Language for the deaf. Although raised in loving, supportive environments, these children were deprived of any exposure to language. Nonetheless, they spontaneously invented a sign language of their own, beginning with single signs at roughly the same age that single words would ordinarily have appeared. And during the following months and years, as they developed more complex sentences, the children produced signs in a serial order according to their semantic role as subject, verb, and object (see Goldin-Meadow 2003 for review).

The songs of sparrows, the calls of monkeys, and the language of human children could hardly be more different, yet they all lead to the same conclusion: each species has a mind of its own that, like its limbs, heart, and other body parts, has evolved innate predispositions that cause it to organize incoming sensations in particular ways. The mind arrives in the world with constraints and biases, "prepared" by evolution to view the world, organize experiences, and generate behavior in its own particular way (Pinker 2002). And because each species is different, the behavior of different species is unlikely to be explained by a few general laws based entirely on experience. Although there may well be some general features of learning that are shared by many species, the behaviorists' principle of equipotentiality ("pigeon, rat, monkey...") is understandable but incorrect.

But what of the behaviorists' second major premise, that the "mind" and "mental states"—if they exist at all—are private and unmeasurable, and cannot be studied scientifically? This view was also challenged, most prominently by the psychologist Edward C. Tolman (1932), who argued that learning is not just a mindless link between stimulus and response. Instead, animals acquire *knowledge* as a result of their experiences.

In 1928, Otto L. Tinklepaugh, a graduate student of Tolman's, began a study of learning in monkeys. His subjects were several macaques who were tested in a room in the psychology department at the University of California at Berkeley (sometimes the tests were held outdoors, on the building's roof, which the monkeys much preferred). In one of Tinklepaugh's most famous experiments, a monkey sat in a chair and watched as a piece of food—either lettuce or banana—was hidden under one of two cups that had been placed on the floor, six feet apart and several feet away. The other cup remained empty. Once the food had been placed under the cup, the monkey was removed from the room for several minutes. Upon his return, he was released from the chair and

7

allowed to choose one of the cups. All of Tinklepaugh's subjects chose the cup hiding the food, though they performed the task with much more enthusiasm when the cup concealed banana.

To illustrate the difference between behaviorist and cognitive theories of learning, pause for a moment to consider the monkey as he waits outside the experimental room after seeing, for example, lettuce placed under the left-hand cup. What has he learned? Most of us would be inclined to say that he has learned that there is lettuce under the left-hand cup. But this was not the behaviorists' explanation. For behaviorists, the reward was not part of the content of learning. Instead, it served simply to reinforce or strengthen the link between a stimulus (the sight of the cup) and a response (looking under). The monkey, behaviorists would say, has learned nothing *about* the hidden food—whether it is lettuce or banana. His knowledge has no content. Instead, the monkey has learned only the stimulus-response associations, "When you're in the room, approach the cup you last looked at" and "When you see the cup, lift it up." Most biologists and laypeople, by contrast, would adopt a more cognitive interpretation: the monkey has learned that the right-hand cup is empty but there is lettuce under the left-hand cup.

To test between these explanations, Tinklepaugh first conducted trials in which the monkey saw lettuce hidden and found lettuce on his return. Here is his summary of the monkey's behavior:

Subject rushes to proper cup and picks it up. Seizes lettuce. Rushes away with lettuce in mouth, paying no attention to other cup or to setting. Time, 3–4 seconds.

Tinklepaugh next conducted trials in which the monkey saw banana hidden under the cup. Now, however, Tinklepaugh replaced the banana with lettuce while the monkey was out of the room. His observations:

Subject rushes to proper cup and picks it up. Extends hand toward lettuce. Stops. Looks around on floor. Looks in, under, around cup. Glances at other cup. Looks back at screen. Looks under and around self. Looks and shrieks at any observer present. Walks away, leaving lettuce untouched on floor. Time, 10–33 seconds.

It is impossible to escape the impression that the duped monkey had acquired *knowledge,* and that as he reached for the cup he had an expectation or belief about what he would find underneath. His shriek reflected his outrage at this egregious betrayal of expectation.

Many years later, Ruth Colwill and Robert Rescorla (1985) carried out a more controlled version of the same experiment. They began by training rats to make two responses, pressing a lever and pulling a chain. When the rats pressed the lever they received a small food pellet; when they pulled the chain they received liquid sucrose. By the behaviorist view, the rats had learned only to press the lever or pull the chain whenever they saw them. By the cognitive view, the rats had formed some kind of mental representation of the relation between a particular act and a specific type of food. To test between these hypotheses, Colwill and Rescorla made either the food pellet or the water unpalatable by adding lithium chloride, a substance that makes rats sick. If the rats had learned which food type was associated with which behavioral act, then those for whom the food pellet had been devalued would avoid the lever but continue to pull the chain, whereas those for whom the water had been devalued would do the opposite. This is exactly what happened.

The results of these experiments challenge the more extreme behaviorists' view that mental states like knowledge, beliefs, or expectations cannot be studied scientifically and may even be an illusion. Instead, they support Tolman's view that learning allows an animal to form a mental representation of its environment. Through learning, animals acquire information about objects, events, and the relation between them. Their knowledge has content, and this content can be studied scientifically.

This conclusion from the laboratory is important, because it encourages us to believe that Darwin was right: we can trace the causation of thought in different species, study its structure, and reconstruct its evolution. But while the scientific study of mind is an exciting prospect, a large dose of humility is in order. For all of their failings, the behaviorists did understand that, whereas behavior can be unambiguously observed and measured, knowledge and the content of mental states are abstract, hard to measure, and difficult even to define. Once you accept the existence of mental states and ascribe causal power to them, you have opened Pandora's box, releasing a host of fundamental questions that are difficult if not impossible to answer.

When we say that a song sparrow's brain "predisposes" it to attend to song sparrow song in a way that it attends to no other, what precisely do we mean? When we claim that a rat has formed an association between bar pressing and a particular type of food, what exactly is the nature of its knowledge? Does the rat think that the bar somehow *stands for* that food? Does it believe that pressing the bar *causes* the food to appear? Can rats distinguish between the relations A *represents* B and A *causes* B? When Pavlov's dog salivated at the sound of a metronome, was this an automatic, unthinking reflex, or did it occur because the metronome brought to mind an image of meat? None of these questions is easy to answer.

Why baboons?

On first—and perhaps even further—inspection, baboons might seem less than ideal subjects for a study of the mind. Among other failings, they are not as closely related to humans as some other nonhuman primates. Baboons are members of the genus *Papio,* Old World monkeys that shared a common ancestor with humans roughly 30 million years ago (Steiper et al. 2004). Baboons are more closely related to humans than monkeys of the New World, but they are much less closely related than the African apes—especially chimpanzees (*Pan troglodytes*)—which diverged from our own ancestors roughly five to seven million years ago. Moreover, the conservation status of baboons confers neither glamour nor prestige on those who study them. Far from being endangered, baboons are one of Africa's most successful species. They flourish throughout the continent, occupying every ecological niche except the Sahara and tropical rain forests. They are quick to exploit campsites and farms and are widely regarded as aggressive, destructive, crop-raiding hooligans. Finally, baboons are not particularly good-looking—many other monkeys are far more photogenic. Indeed, through the ages baboons have evoked as much (if not more) repulsion than admiration.

Baboons are interesting, however, from a social perspective. Their groups number up to 100 individuals and are therefore considerably larger than most chimpanzee communities. Each animal maintains a complex network of social relationships with relatives and nonrelatives—relationships that are simultaneously cooperative and competitive. Navigating through this network would seem to require sophisticated social knowledge and skills. Moreover, the challenges that baboons confront are not just social but also ecological. Food must be found and defended, predators evaded and sometimes attacked. Studies of baboons in the wild, therefore, allow us to examine how an individual's behavior affects her survival and reproduction. They also allow us to study social cognition in the absence of human training, in the social and ecological contexts in which it evolved.

In Darwin's theory of evolution by natural selection, necessity is the mother of invention. Traits arise or are maintained because they help the individuals who possess them to solve a problem, thereby giving those individuals an advantage over others in survival and reproduction. A blunt, heavy beak allows a finch to crush hard, dry seeds and survive a withering dry season; antlers enable a stag to defeat his rivals and mate with more females. The finch's beak and the stag's antlers did not arise at random; they evolved and spread because of their adaptive value. To understand the evolution of a trait, therefore, we need to understand how it works, and what it allows an individual to do that might otherwise be impossible.

And brains, Darwin realized, were biological traits like any other. To understand how they evolved, we must understand the problems they were designed to solve. In recent years, studies of the brain, intelligence, and evolution in animals have produced two general conclusions that will guide our study of baboon metaphysics.

First, natural selection often creates brains that are highly specialized. Arctic terns (*Sterna paradisaea*) migrate each year from one end of the earth to another, *Cataglyphis* ants navigate across the featureless Sahara, bees dance to signal the location of food, and Clark's nutcrackers (*Nucifraga columbiana,* a member of the crow, or corvid, family) store and recover tens of thousands of seeds during the fall and winter. Yet despite these specialized skills, there is no evidence that terns, ants, bees, or nutcrackers are generally more intelligent than other species. Instead, they are more like nature's idiots savants: brilliant when it comes to solving a specific, narrowly defined problem, but pretty much average in other domains.

Specialized intelligence may be widespread in animals because brain tissue is costly to develop and maintain. The human brain uses energy at a rate comparable to that used by the leg muscles of a marathon runner when running (Attwell and Laughlin 2001). If brain tissue is energetically expensive, the cheapest way to evolve a specialized skill may be through a small number of especially dedicated brain cells rather than a larger, general-purpose brain. For arctic terns, the ability to fly from pole to pole in the spring and fall is adaptive because it allows the birds to live in perpetual summer. As a result, selection has favored individuals with the neural tissue needed to navigate great distances using the sun, the stars, and the earth's magnetic field. But it has done so in the cheapest, most energy-efficient way possible—by selecting specifically for navigational skills.

The second general conclusion to emerge from recent research is that the domain of expertise for baboons—and indeed for all monkeys and apes—is social life. Most baboons live in multimale, multifemale groups that typically include eight or nine matrilineal families, a linear dominance hierarchy of males that changes often, and a linear hierarchy of females and their offspring that can be stable for generations. Daily life in a baboon group includes small-scale alliances that may involve only three individuals and occasional large-scale, familial battles that involve all of the members of three or four matrilines. Males and females can form short-term bonds that lead to reproduction, or longer-term friendships that lead to cooperative child rearing. The result of all this social intrigue is a kind of Jane Austen melodrama, in which each individual must predict the behavior of others and form those relationships that return the greatest benefit. These are the problems that the baboon mind must solve, and this is the environment in which it has evolved.

Social problems, of course, are not the only challenges. Baboons also need to solve ecological problems, like finding food and avoiding predators. But these problems are also overwhelmingly social. One of the most difficult aspects of finding food arises from the fact that as many as 100 other individuals in your group also want the food for themselves. And the best way to avoid being taken by lions, leopards, crocodiles, or pythons is to live in a group, with all of the opportunities and compromises that group life entails. Any way you look at it, most of the problems facing baboons can be expressed in two words: other baboons.

The study group and data collection

The focus of our research is a group of chacma baboons (*Papio hamadryas ursinus*) living in the Moremi Game Reserve in the Okavango Delta of Botswana. We began our study in 1992, but before our arrival the group had been observed more or less continuously for 14 years by W. J. Hamilton III and his students at the University of California at Davis. Because the baboons have endured interlopers for three decades, they are completely habituated to humans walking among them and tolerate our presence with diffident aplomb, if not affection. Even the oldest female in the group, the curmudgeonly and mean-spirited Sylvia, has had to put up with human observers since her birth in 1982. Between 1992 and 2006, group size averaged 80 individuals, with fluctuations depending on rates of infanticide, predation, and male immigration. The number of adult females has varied from 18 to 28 and the number of adult males from 3 to 12.

When following the baboons, we and our colleagues collect three sorts of data. First, each day we note all demographic changes in the group, including births, deaths, immigrations, emigrations, and sexual consortships. Second, we conduct 10 minute-long "focal animal samples" (Altmann 1974) on each individual following a systematic rotation. These samples supply us with a continuous record of the baboons' interactions and social partners and provide the data to document the continuous soap opera that constitutes baboon life. We also note specific other events—like fights, alliances, interactions between groups, and encounters with predators—on an ad libitum basis, whenever they occur. Third, we make audio recordings of the baboons' vocalizations, for both acoustical analysis and "playback" experiments. We describe these experiments in detail in Chapters 5 and 6. Finally, between 2002 and 2005 we have collected weekly fecal samples from all adult males and females for the extraction of testosterone (from males) and glucocorticoids (from males and females). Glucocorticoids are a class of steroid hormone associated with stress.

The beauty of a fecal sample—if that is the appropriate term—is that it allows us to measure a biological response that cannot be observed. It can also be collected without itself inducing stress, as would certainly happen if we tried to extract blood. The data from fecal samples allow us to look beneath the surface of baboon society and ask, "Who is under stress? Why? And how it is alleviated?" Like humans, baboons have families, seek mates, form friendships, and suffer fear and anxiety from events both social and environmental. Unlike humans, though, baboons cannot explain the causes of their anxiety to us; indeed, as we will see, they may not even be explicitly aware of feeling anxious or depressed. Like their behavior and vocalizations, the baboons' hormonal profiles allow us to ask them, indirectly, what they think and how they feel.

How this book is organized

In writing this book, we had to decide whether to include material from our earlier book on vervet monkeys (*Cercopithecus aethiops*), *How Monkeys See the World* (Cheney and Seyfarth 1990). We knew that we could not operate under the conceit that our readers would remember anything from that volume, but at the same time we wanted to avoid *The Bride of How Monkeys See the World*. We also had to resolve how exhaustively we would review the vast literature on animal cognition. In the end, we decided that we would focus primarily on research that was directly relevant to our work on baboons. We therefore discuss vervet monkeys only sparingly and make no attempt to consider, for instance, whether animals have "cognitive maps" of their environment, can represent numerical quantities, or make optimal foraging decisions. This is not due to laziness, nor is it because we believe that baboons simply stumble about their habitat with no inkling about where they are, where they are going, or what they are eating. Instead, we avoid these and many other important questions because we were unable to investigate them directly (two good reviews of animal cognition are Shettleworth 1998 and Tomasello and Call 1997).

The link between primates' intelligence and the complexity of their social behavior may seem obvious, but this has not always been the case. In Chapter 2, we take a historical perspective and examine a curious fact about our ancestors' knowledge of their closest animal relatives. For centuries people have known that, of all the creatures in the world, monkeys and apes are most like us. Ironically, however, scholars reached this conclusion without knowing anything at all about the characteristics of primates that make them most human: their social life. Because Western scientists learned about primates by examining corpses or observing single animals brought home as pets, few if any ever learned what can be discovered only through long, patient observation: that the most human features of monkeys and apes lie not their physical appearance but in their social relationships.

In Chapter 3 we describe the ecological setting in which our work takes place and the predators that so affect baboons' lives. In Chapters 4 and 5 we introduce the protagonists with a discussion of social behavior and life histories among males (Chapter 4) and females (Chapter 5), in all of their familial complexity, friendships, alliances, stress, and Machiavellian intrigue. As part of this description we introduce, in Chapter 5, the method of field "playback" experiments that we use to explore what baboons know about the relations that exist among others. In doing so, we present one of our central arguments—that even though baboons lack language, their vocal communication is rich enough in meaning to tell us a great deal about how they think. Primate vocalizations, in fact, provide the key that unlocks the primate mind.

Whereas Chapters 1 through 5 are introductory, historical, and descriptive—designed to introduce readers unfamiliar with baboons to the monkeys' habitat, behavior, and social structure—Chapters 6 through 11 delve more deeply into the scientific questions that guide our research. In Chapter 6 we describe experiments designed to test baboons' knowledge of their social companions. The results show that baboons are good psychologists: they recognize their companions as individuals, observe their behavior, and create, in their minds, a hierarchical representation of society based on matrilineal kinship and dominance rank. The social knowledge of baboons is too varied and complex to be explained by simple learning mechanisms. Instead, we propose that natural selection has led to the evolution of a mind innately predisposed to search for the patterns and rules that underlie other baboons' behavior.

In Chapter 7 we examine baboons' knowledge of their companions in light of the "social intelligence" hypothesis, which argues that the demands of living in large social groups have placed strong selective pressure on the evolution of the primate mind. The average value for relative brain size in primates exceeds the average value for other mammals, and primate brains contain many areas specialized for dealing with social stimuli. Baboons and other monkeys recognize each other's ranks and kin relations, and their reproductive success and ability to overcome stress depend on their skill in forming social relations. Similar social skills, however, are also found in nonprimate species that live in large social groups, including dolphins, hyenas, and pinyon jays. Furthermore, even relatively asocial species appear to monitor other individuals' social interactions. It therefore remains unclear whether social intelligence in animals depends on taxonomic affiliation, group size, or some other combination of factors.

In *How Monkeys See the World,* we concluded that, for all their intriguing similarities, the societies of nonhuman primates were fundamentally different from our own because monkeys and apes lack a "theory of mind"—the ability to attribute mental states like knowledge and belief to others. In Chapter 8 we reconsider this conclusion in light of experiments conducted over the past 15 years by ourselves and many others. In Chapter 9 we consider the related question of whether baboons or any other primates are aware of their own mental state—that is, whether they have anything like our concept of self.

We take it for granted that human words express thoughts and that language provides a window onto the mind. Surprisingly, however, few people have ever applied this idea to animals. In Chapter 10 we review what is known about the vocal communication of baboons and confront directly one of the questions that behaviorists—perhaps wisely—avoided: What does one baboon's vocalization "mean" to another? We also consider the complex relation between language and thought, but from a perspective not usually found among those who work exclusively on humans: we ask what thought is like in a creature without language. In Chapter 11 we consider what our work has to say about the evolution of language. Finally, in Chapter 12 we return to the challenge posed by Darwin's famous quotation—that an understanding of baboon metaphysics can shed light on the evolution of human mind and behavior.

TWO

The Primate Mind in Myth and Legend

Our descent, then, is in the origin of our evil passions!! The Devil under form of Baboon is our grandfather! CHARLES DARWIN, 1838: *NOTEBOOK M*

[Ahla, the baboon] is not only eager but really a maniac when it comes to putting back the lambs with the mothers. She can't wait until the door between the two enclosures is opened.
WALTER HOESCH, 1961: *ON GOAT-HERDING BABOONS*

The baboon in Egypt: god, scribe, and policeman

Baboons range widely throughout the African continent, so it is perhaps not surprising that they appear often in ancient Egyptian mythology and art. Beginning in at least the fourth millennium B.C., baboons were associated with the underworld and considered to be embodiments of the dead, no doubt in part because they resembled humans so closely. The word "baboon" may derive from the baboon god Baba, or Babi, a supernaturally aggressive deity who was revered during the Predynastic Period. Perhaps because the ancient Egyptians could not help but notice male baboons' sexual zeal and prominent genitalia, the baboon god ensured that the dead would not suffer from impotence in the afterlife. Indeed, baboon feces were used as an ingredient in aphrodisiacs.

By the time of the Old Kingdom, around 2400 B.C., baboons had become associated with Thoth, the god of wisdom, science, writing, and measurement. In tomb paintings and sculptures, baboons instructed scribes in their tasks and weighed the hearts of the deceased. Baboons also came to be identified with the sun god Re, probably because the loud dawn choruses of male baboons' *wahoo* calls were taken as a sign that they were worshiping the sun. In addition to associating with Thoth, baboons were believed to stand by Re in his boat as he traveled across the sky. Even into the late Ptolemaic periods, baboons were still regarded as sufficiently sacred to be mummified and kept in colonies at temples (Budge 1969; David 1998; Redford 2002).

But the ancient Egyptians did not just depict baboons as deities. They also portrayed them in many other guises, not only as scribes but also as musicians, sailors, shipwrights, fishermen, and even vintners. Most, if not all, of these depictions were doubtless fanciful—it seems unlikely that baboons ever tended grapes or built ships. More credibly, baboons were depicted as captives brought from the south, as pets on leashes, or as dancers or jesters. Some paintings show them climbing trees to collect figs and dates for their master, and—even more plausibly—pilfering fruit from baskets (Wilkinson 1879; Janssen and Janssen 1989). Hieroglyphics from tombs of the New Kingdom accompany some of these pictures with remarks like "A monkey carries a stick (for dancing), though its mother did not carry it," suggesting the artist's appreciation for the baboon's ability to learn. Other hieroglyphics comment on baboons' capacity to understand words (Janssen and Janssen 1989). Baboons even appear in the role of police assistants. One illustration from the Old Kingdom mastaba of Tepemankh at Saqarra shows two baboons on leashes—one a female carrying an infant and the other an adult male—grabbing thieves in the market place (Fig. 1). The accompanying

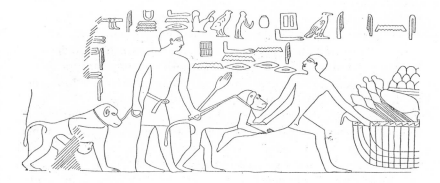

Figure 1. An Egyptian tomb painting from the Old Kingdom depicts two baboons acting as police assistants, attacking a thief in the marketplace.

hieroglyphic reads, "Fear for this baboon" (Smith 1946; Janssen and Janssen 1989).

The Egyptians probably derived much of their knowledge about baboons from pets, temple colonies, and stories emanating from Nubia or other remote areas. Although most of these portrayals probably served a religious or humorous function, they also show that the Egyptians were not entirely ignorant of baboons' natural social behavior: male baboons do, for example, participate in loud *wahoo* contests in the early morning. If any early Egyptian ever did take the time to observe the baboons' natural social interactions, however, he left no record of his observations.

European and Japanese attitudes

No clear record exists of the first contact between a European scientist and a nonhuman primate (Janson 1952). As far as we can tell from Aristotle's *Historia Animalium* and other Classical texts, the first primatological subjects to be studied by western scientists were baboons and the Barbary macaques (*Macaca sylvanus*) that inhabited the southern shores of the Mediterranean (Spencer 1995). These animals came to scientists either as corpses to be dissected or as single animals to be held in cages and observed.

Classical speculation about the mind and behavior of primates was part of a more general curiosity about all animals, and about the fundamental differences between animals and human beings. Aristotle believed that, when it came to emotions, the difference between animals and humans was only a matter of degree. In both humans and animals, tameness graded into wildness, docility into stubbornness, boldness into cowardice or fear, and confidence into anger (Sorabji 1993). By contrast, in matters of intellect Aristotle drew a sharp distinction between humans and all other creatures, including the nonhuman primates. Unlike humans, he argued, animals were completely lacking in reason, intellect, thought, belief, and, as a consequence, language. How, then, did they manage to deal with the world? To make up for their lack of intellect, Aristotle argued, animals have an elaborate, expanded, but intellectually limited, form of perception (Sorabji 1993). Dogs (*Canis familiaris*) are extremely skilled at identifying and tracking scents, but they know only how to detect and react to an olfactory stimulus. They have no true knowledge or beliefs about scents, nor about the causal relations that link a particular scent with its owner. Dogs, in Aristotle's view, just react; humans understand.

Scapegoat and trickster

Were monkeys and apes any different? The Greeks and Romans recognized clearly that, among all animals, nonhuman primates were the creatures most similar to human beings. But their anatomy did not elevate their status; instead, quite the reverse occurred. Convinced, like Aristotle, that all animals were fundamentally different from humans on intellectual grounds, Classical scholars ignored both the anatomical evidence and Aristotle's argument for continuity in emotions. Instead, they adopted a kind of "reverse Darwinism" in which the more an animal resembled a human, the more it was shunned, made into an object of ridicule, and declared to be fundamentally different. The general view is summarized by the dictum of the Roman poet Quintus Ennius: *"Simia quam similis turpissima bestia nobis"* ("How like us is that ugly brute, the ape").[1] As the art historian H. W. Janson points out, "the ape was *turpissima bestia* precisely because it was *quam similis nobis*. As an unworthy pretender to human status, a grotesque caricature of man, the ape became the prototype of the trickster, the sycophant, the hypocrite, [and] the coward," as well as the symbol of extreme physical ugliness. Or as Plato put it, "The most beautiful of apes is ugly compared to man and the wisest of men is an ape beside God" (Janson 1952:14–15; McDermott 1938; Corbey 2005).

In Japan, where humans have coexisted with Japanese macaques (*Macaca fuscata*) for thousands of years, monkeys played a similarly ambiguous role in everyday life and legend. In Japanese legends monkeys were often portrayed as foolish and vain creatures whose servitude to their master gods eventually earned them courage, generosity, wisdom, and loyalty. Many of these legends arrived with Buddhism from India, where the monkey god Hanuman is still revered as a loyal and intelligent servant to the mythical King Rama. The Japanese recognized that monkeys were obviously the most humanlike of all animals: even today, the monkey is the only creature referred to in the Japanese language by the term *san,* the address form used for humans. As a "special" animal, monkeys were regarded as mediators between humans and deities and were thought to have the power to maintain the health and

---

1. In early writings, no distinction was made between monkeys and apes.

cure the illnesses of horses. Monkey parts were used to cure human illnesses and bring good luck. On the other hand, the monkey's elevated position coexisted uneasily with its image as a false pretender to human status, a scapegoat, a charlatan, and a harbinger of bad luck. Before the 13th century, portraits of the monkey as a semi-god predominated; in the art of the next 600 years, however, its role as an unlucky trickster took over (Ohnuki-Tierney 1987). For the Japanese, as for the Greeks and Romans, the image of nonhuman primates as pretenders to human status seems to have been crucial. The animals' physical similarity to humans was both their salvation and their curse.

Ironically, there was in this reverse Darwinism the germ of an evolutionary theory linking humans with monkeys and apes. Because these animals looked so much like humans, the Greeks, Romans, and Japanese all knew that they were somehow involved in fundamental questions about the origin of human beings. The Greeks believed that humans had originally lived in caves, freely mingling with animals and in particular with apes, until with the aid of the gods they gradually acquired civilization (Boas 1948; Janson 1952). The Japanese had similar origin myths. But the revulsion that arose in each of these cultures whenever monkeys and apes were compared to humans apparently prevented scholars from recognizing that their own legends might be more than just mythical accounts.

The image of monkeys and apes as humans manqué is perhaps nowhere better expressed than in a Jewish legend about the fall of the Tower of Babel. The story of the tower is of course well known: according to the Old Testament, the people of the earth had at that time only one language. Jealous of God's power, they gathered together and resolved to build a tower that would allow them to ascend into heaven. God saw them doing this, and realized that if they succeeded nothing could ever again be denied to them and they would become all-powerful. So to divide and weaken the people, God not only destroyed their tower and scattered them throughout the earth but also confounded their language, making their speech mutually unintelligible so that they would never again be able to communicate with one another.

But in Jewish legend it was not the humans who received the worst of God's wrath. There were also at the time "apes and monkeys" at work on the tower, and these primates were involved in a project that outraged God even more than the construction of the tower itself: they had begun to build idols that would stand atop the tower and be worshipped, in place of God, by all of the creatures on Earth. God punished the monkeys and apes by taking away their language altogether and banishing them into the forest (Janson 1952; Ginsberg 1968).

Emotions, impulses, and lust

Just as in 13th-century Japan the image of the monkey changed from healer and sacred messenger to scapegoat and trickster, so in the European Middle Ages the image of monkeys and apes took a decided turn for the worse. Their downfall coincided with the rise of Christianity (Janson 1952). Western theologians now used monkeys and apes as living examples of what man would become if he turned away from God and gave way to his baser instincts. Monkeys and apes were no longer just devious sycophants—they now became creatures completely at the mercy of emotions, sadistic impulses, and lust. In the most extreme characterizations, they were depicted as the devil himself (or at least the devil's agents), sent from out of the land of darkness (usually Egypt) to perform Satan's work and lead people—particularly women—into sin.

Baboons were the object of particular revulsion and scorn, in large part because of their supposed deviant sexual appetites. Writing about a captive baboon in Paris in 1775, Buffon remarked that "the baboon was insolently lascivious, and satisfied its strong desire in public. It seemed also to make a parade of its nakedness, presenting its posteriors oftener to the spectators than its head; but it was particularly impudent in the presence of women, and plainly showed its immoderate desires before them by an inexpressible lasciviry."

In 1699 the English anatomist Edward Tyson published a monograph on the anatomy of a juvenile ape that had died shortly after its arrival in London from Angola (Spencer 1995). The monograph was important for two reasons. First, it was the first scientific text to draw a clear distinction between the groups we now know as monkeys (particularly Old World monkeys) and apes. Although Tyson identified his primary specimen incorrectly (he called it an "orang-outang," *Pongo pygmaeus,* but coming from Angola it must have been either a chimpanzee, *Pan troglodytes,* or a gorilla, *Gorilla gorilla*), he nonetheless recognized that it was much more humanlike than many other "apes" (i.e., monkeys) that had been formerly described, and he concluded that the primates of the Old World must fall into two groups, one more closely related to humans than the other.

Second, Tyson's monograph spurred a new wave of popular speculation about the exact differences between humans and their closest animal relatives. The speculation was born of uncertainty about what his creature actually was. According to Tyson's anatomical work, the supposed orangutan was clearly not human. But reading between the lines it was equally clear that Tyson believed the difference was very slight: he even called his specimen *Homo sylvestris*. What should be made of this chimera? Doubt, wonderment, and outrage expressed themselves in satire.

The satirical accounts came in two forms. In the most common genre, an ape was introduced into polite society, where his social and intellectual successes—along with his curious failures—served to show just how far education and proper training could subdue his animal nature and elevate him to full human status. In 1732, Alexander Pope, John Arbuthnot, and other prominent authors published the *Essay of the Learned Martinus Scriblerus, Concerning the Origin of the Sciences,* a collaborative work of the Scriblerus Club, designed to satirize the excesses of erudition (Ashley Montagu 1941; Janson 1952). In it they locate the beginnings of art and science in the work of monkeys and apes. Among their most advanced culture heroes is "Orang Outang the great, whose unhappy chance it was to fall into the hands of the Europeans [and] whose value was not [formerly] known to us, for he was a mute philosopher." Like the authors of the biblical legends concerning apes and the Tower of Babel, the members of the Scriblerus Club concluded that Orang the Great could think, write, and reason philosophically, but not speak. Thus it was language—or at least the ability to engage in speech—that separated *Homo sapiens* from *Homo sylvestris.*

Nearly a century later, in 1817, Thomas Love Peacock echoed this theme in *Melincourt,* the tale of a young orangutan from Africa who distinguishes himself in English society as Sir Oran Haut-ton. Sir Oran is regarded by his human companions as superior to most of their compatriots in gallantry and nobility of feeling, largely because he rescues a maiden in distress without taking advantage of her. Most of the time he is able to control his animal impulses, but on one occasion, after drinking wine, he leaps out of a window and "goes dancing along the woods like a harlequin." Although he cannot speak, Sir Oran is nonetheless elected to Parliament, where his lack of speech is an asset rather than a hindrance because it gives him the reputation of a powerful but cautious thinker (Henkin 1940).

The second genre of satirical account imagined a traveler in some remote corner of the world who discovers an animal society and then returns to describe what he has seen among these bizarre creatures. Jonathan Swift's Gulliver is, of course, the most well known of these philosophical explorers, the forerunners of modern ethologists. When

Gulliver encounters the Yahoos, he is shocked to find that they resemble humans in every detail of their bodies yet they behave like animals. They have no language, but can only bellow loudly and repeatedly. By contrast, the Yahoos' masters, the noble and generous Houyhnhnms, have both the ability to reason and the gift of language even though their bodies are those of horses. At first the Houyhnhnms assume that Gulliver is himself a Yahoo, but once he demonstrates his good manners, cleanliness, and linguistic ability they treat him more like an equal (Swift 1726).

The satirical accounts of the 18th and 19th centuries mocked not only the apes themselves but also the scientists and philosophers who were so undecided about the animals' classification. And the satirists had a point: at a time when zoological taxonomy was undergoing a revolution, no one knew exactly what to do about these creatures that were so much like us and yet so obviously different. In the first edition of his *Systema Naturae* (1735), Carolus Linnaeus, the father of modern taxonomy, based his classification of mammals exclusively on anatomical characteristics and lumped the apes as they were then known together with humans in the group Anthropomorpha. This provoked an outraged response from—among many others—the French naturalist Georges-Louis Leclerc, comte de Buffon, who strongly objected to Linnaeus' exclusive reliance on anatomical features. Like his countryman Rene Descartes, Buffon believed that the possession of reason and language distinguished humans from all other animals, and that to ignore the dualistic nature of the human condition was to ignore the noblest feature of our species (Spencer 1995; Corbey 2005). Twenty-three years later, in the 1758 edition of the *Systema,* Linnaeus responded to his critics by separating the various members of the genus Homo from all other animals and basing his classification not only on anatomical features but also on temperament, character, type of garments worn (if any), and forms of government.

Looking back, three characteristics about peoples' views of monkeys and apes from Classical times to the present are striking. First, reverse Darwinism appears in every age and every culture. Throughout time and around the world, people have readily accepted the idea that two animal species with similar morphology must bear some close genealogical relationship to one another, but they have balked and indeed reversed this rule when the species in question are human and nonhuman primates. Even today, the Mende hunters of the Ivory Coast believe that chimpanzees possess near-human levels of society and culture but argue that it is morally and physically dangerous for humans and chimpanzees to live in the same environment because the chimpanzees "set low standards of behavior to which humans may be tempted to descend" (Richards 1995). Similarly, in the United States, "microevolution" is now widely, if reluctantly, accepted because of irrefutable facts like the emergence of bacterial strains that are resistant to antibiotics. However, the majority of Americans continue to deny that humans and apes share a recent common ancestry.

Second, in order to defend a strict, dichotomous division between themselves and their closest animal relatives, humans have needed to come up with a crucial, defining criterion: something important that we clearly have and nonhuman primates lack. For most of the past 2000 years, language has remained the preferred choice. This may seem obvious today, when debates about the humanness of apes (or the animal nature of humans) inevitably come down to language, but it has not always been the case. At various times in the past, the sine qua non of humanity has been our ability to overcome our baser instincts, our hairlessness, our upright posture, our large brain size, or our ability to make tools. As recently as 1986, the country music singer Dolly Parton suggested that "what separates us from the beasts is our ability to accessorize." Language, then, has not always been viewed as the defining feature that sets us apart from apes; it is, however, the feature that has stood the test of time.

Finally, as we look back on earlier views of human and nonhuman primates, we are struck not just by the attitudes themselves but even more forcefully by the biased samples on which these attitudes were based. For centuries, scientists and philosophers formulated their views of monkeys and apes entirely on the basis of either dissections of dead specimens, observations of lone, captive individuals, or (as in the case of Richard Jobson [1623]) glimpses of an entire group seen from a distance. Their judgments, as a result, were made without any awareness of the very traits—like systems of kinship and dominance, or complex social alliances—that today make monkeys and apes seem most like human beings.

Modern studies of monkeys and apes

The historical bias against studying nonhuman primates in natural, social settings seems all the more odd because people have always known that monkeys and apes are fundamentally social creatures. In Japan, hunters believed so strongly that the macaque is a group-living animal that they held a taboo against killing any monkey found on its own (Ohnuki-Tierney 1987). Despite this enlightened view, Japanese opinions about the minds and behavior of nonhuman primates were derived almost exclusively from what they knew about monkeys' performances with humans (Asquith 1995). Although monkeys range freely on many of Japan's islands, Japanese artists, like those of India and Western Europe, have usually depicted them amidst humans, far from their natural habitats and social groups (Janson 1952; Ohnuki-Tierney 1987; Asquith 1995). As a result, it was not until the 1950s that Japanese scientists—and through them the rest of the world—learned that the society of Japanese macaques is organized around a ranked group of matrilineal families (Kawai 1958; Kawamura 1958).

Similarly, the Mende hunters of the Ivory Coast claim to have a sophisticated knowledge of chimpanzee behavior, and in fact do know a great deal about the chimpanzees in their area, including their use of tools to crack open nuts and plants for self-medication (Richards 1995). But the Mende's knowledge does not extend to the details of chimpanzee society. They are unaware, for example, that the core of a chimpanzee community is a group of long-term resident males, some of whom may be related to one another, and unrelated immigrant females, or that within each community individuals come and go in fluid, transient subgroups (Goodall 1968).

In Western Europe and America, where there are no indigenous nonhuman primates, the inclination to study monkeys and apes within the context of human society has been even more apparent. In the early 20th century, for instance, French scientists recommended the establishment of a model village in French Guinea that would serve as a training ground for the civilization of wild apes. Native women would act as nurses and guides. British scientists planned a monkey college to make chimpanzees human, and German scientists established a colony in North Africa to study how chimpanzees solve problems (Harraway 1989). In each case, Western scientists made the effort to travel to the apes' native habitat in Africa, but once there put all notion of naturalistic research aside. Later in the 20th century, the American psychologists Robert Yerkes and Harry Harlow embarked on major studies of chimpanzee and rhesus macaque behavior, but again focused their efforts exclusively on captive individuals, typically housed alone in highly restricted environments.

The twin beliefs that language constitutes the crucial difference between humans and nonhuman primates and that these animals are best studied in captivity are most evident in the "ape language" projects, which began haltingly in the 1930s, flourished through the 1960s and '70s, and continue to this day. Like Classical and medieval scholars, the scientists who conducted these studies have been fascinated by the idea that language is what makes us different from all other creatures, and they have accepted, either explicitly or implicitly, the view that language and thought in apes cannot be studied in the apes' own societies but must instead be explored by bringing them into ours and teaching them to communicate like us.

But the artificial settings of the ape-language projects have made their results difficult to interpret. When an ape has failed to achieve some linguistic milestone—for example, when the chimpanzee Nim Chimpsky learned words but was unable to combine them into sentences—critics have charged that the animal's failure did not reflect his true ability. Instead, Nim failed because he had an abnormal upbringing, his training was nothing more than mind-numbing repetition, and he was tested under circumstances that were artificial, contrived, and entirely too rigid. Nim and other captive apes would have done better, the critics have argued, if their experience as youngsters, like the experience of human children, had been more conducive to learning language (Terrace 1979; Seidenberg and Pettito 1979). Conversely, when captive apes have apparently succeeded in acquiring a linguistic skill— for example, when the bonobo Kanzi learned to respond correctly to complex sentences like "Put the ball in the basket"—critics have argued that such achievements are anecdotal, and in any case do not reflect a genuine understanding of grammatical rules. Kanzi, the critics have argued, may simply have recognized "put," "ball," and "basket" as separate signs and put the ball in the basket because the basket could not be put in the ball (Savage-Rumbaugh 1986; Wallman 1992). Small wonder that the scientists involved in these studies have often felt trapped in a game of "heads you win, tails I lose."

There is no doubt that the best laboratory experiments achieve a level of precision and control that field research on natural groups of monkeys and apes cannot begin to match. Many captive studies are also highly creative and have succeeded in documenting abilities that can only be hinted at by field observations. We discuss some of these experiments in Chapters 7 through 10. At the same time, however, laboratory experiments can illuminate a species' abilities only if their results can be placed within the context of an animal's natural social behavior. In the absence of such grounding, they remain difficult, if not impossible, to interpret.

Studies of captive apes may be particularly difficult to interpret because human "enculturation" may affect the apes' cognitive abilities and performance on tests. To date, most of the evidence that chimpanzees' cognitive abilities are superior to those of monkeys comes from chimpanzees that have had prolonged contact and/or training with humans. But there may be as many differences between the performance of "enculturated" and "natural" chimpanzees as there are between apes and monkeys generally. In one experiment specifically designed to test the effect of human enculturation, Tomasello and colleagues (1993) compared the imitative abilities of chimpanzees raised by humans (but not language-trained), chimpanzees raised by their own mothers, and two year-old children. Each subject was shown a number of novel actions and scored according to whether the subject imitated the action of the demonstrator. Mother-reared chimpanzees did not, whereas human-reared chimpanzees and children did. In another experiment (Carpenter et al. 1995), investigators examined the use of joint attention by chimpanzees and children when learning to imitate a task involving novel objects. Again, children and enculturated chimpanzees looked back and forth from the object to the demonstrator and used gestures to direct the demonstrator's attention, whereas mother-raised chimpanzees did not.

Does exposure to humans somehow enhance chimpanzees' cognitive capacities? Human trainers actively engage their chimpanzee subjects' attention when interacting with them or instructing them in the use of signs. As a result, these apes may come to view humans as intentional agents who have goals and motives (Tomasello and Call 1997). Alternatively, through exposure to humans chimpanzees may become familiar with human artifacts and training regimes, which in turn facilitates learning. Whatever the explanation, evidence that exposure to humans affects chimpanzees' cognition complicates any comparisons between the apes' performance in captivity and their behavior in the wild.

Baboons in southern Africa

Farmers in southern Africa will tell you that they hate baboons. Baboons raid their orchards, decimate their cornfields, destroy their irrigation pipes, pollute their water tanks, and kill their sheep and goats. In response, the farmers shoot baboons whenever they get the chance.

But the farmer-baboon relationship is not as simple as it seems. Long before any Western scientist had begun to study monkeys in the wild, southern African farmers had gained a grudging appreciation not just of baboons' intelligence but also of their society.

The farmers' knowledge may have arisen from the peculiar ecology of their farms. In much of southern Africa, baboons sleep at night and rest during the day on steep, rocky cliffs, where they are safe from leopards and can find comfortable, baboon-sized sitting-places to groom and play in the shade. The farms lie in the valleys below. The result is a pastoral landscape in which farmers toil in the fields while groups of baboons sit like vultures on the cliffs, grooming, playing, watching, and waiting for just the right moment to raid a cornfield or steal a baby goat. And if a farmer looks up, he sees—albeit at a distance—what few European or American scientists ever observed: an entire baboon group, going about its daily activities. He sees infants and juveniles gathered around their mothers, females grooming one another, and males giving loud alarm barks if a human begins to approach. Observing these behaviors, a curious farmer might be prompted to ask, "Do they have families? Do males protect their offspring?"

Eugene Marais (1872–1936) was raised on a farm near Pretoria, South Africa, "as completely cut off from the rest of the civilized world as the loneliest isle in the Pacific" (cited in Ardrey 1969). But despite its remote location, the farm undoubtedly contained baboons. After receiving his education, Marais began work as a journalist, eventually becoming the editor of a small, rural, Afrikaner paper. He wrote well in both English and Afrikaans, but confined his writing to the latter language for patriotic reasons. A few years later, when his wife died in childbirth, Marais went to London, where he was admitted to the bar. When the Boer War began, he returned to South Africa to support his people in their fight against the English.

The Afrikaners' defeat plunged Marais into deep depression, and sometime around 1903 or 1904 he retreated to a small farm in the Waterburg district to recover. Farmers had abandoned this area during the war, leaving baboons free to forage wherever they wished. Even though the farmers were now beginning to return, they had no guns to drive the simian raiders away. The baboons were bold, unafraid of humans, and Marais was able to watch them at close range. In a letter written in 1935 he stated:

In other countries you are lucky if you catch a glimpse of the same troop twice in a day. I lived among a troop of wild baboons for three years; I followed them on their daily excursions; slept among them; fed them night and morning on mealies; learned to know each one individually; ... . I have an entirely new explanation of the so-called subconscious mind and the reason for its survival in man. I think I can prove that Freud's entire conception is based on a fabric of fallacy. No man can ever attain to anywhere near a true conception of the subconscious in man who does not know the primates under natural conditions. (cited in Ardrey 1969:20)

Today, Marais' books on baboons, *The Soul of the Ape* (1922) and a series of articles published as *My Friends the Baboons* (1939), seem rambling, idiosyncratic, and anthropomorphic in the extreme. But Marais was a scientific pioneer because he realized that the essence of baboon life was their society. His insights resulted from a lucky confluence of events: his background in rural South Africa, where he encountered baboons in groups; his luck in the Waterburg, where he could observe them at close range; and his need to escape the world and recover from deep depression. This last bit of fortune—if you can call it that—gave him the time to study animals in depth, the motivation to recognize them as individuals, and a deep, thoughtful curiosity about the subconscious and the relation between individuals and society.

Oxcart drivers, signalmen, and shepherds

One consequence of the war between farmers and baboons was that, inevitably, some infant baboons were orphaned. And for perhaps the first time since the ancient civilizations of Egypt, some of these orphans were adopted by local people and put to work, in a few celebrated cases as oxcart drivers, railway laborers, and goatherds on farms. Particularly in their role as "herdsmen," these baboons displayed the kind of sophisticated social knowledge that we know today is the hallmark of baboon metaphysics. In all of their assigned roles, baboons served their masters through their intelligence. The story of Jack the Signalman provides one such example.

During the latter part of the 1800s, the Cape Government Railways opened the first line from Cape Town to Port Elizabeth. In the inland town of Uitenhage, a railway guard named James Wide earned the nickname "Jumper" because of his skill in leaping from one moving train to another. Alas, the inevitable happened; one day Jumper fell. The train ran over him and both legs had to be amputated at the knees. In a desperate attempt to keep his job, Jumper made himself a pair of pegged legs by strapping pieces of wood to his lower body. He also built a hand trolley that made him more mobile. The company agreed to hire Jumper as a signalman, but his work remained a struggle.

One Saturday morning in the Uitenhage marketplace, Jumper noticed an ox wagon being driven into town by a young baboon, who acted as "voorloper" (oxen leader). Convinced that such an intelligent creature might be useful to him, Jumper persuaded the owner to give up his favorite pet. The baboon was called Jack. In parting, the owner warned Jumper that every evening Jack should be given "a tot of good Cape brandy"; otherwise he would spend the next day sulking and refuse to work.

Jack soon learned to drive Jumper to work each morning. He pushed the trolley on the uphill grades, then leapt aboard in great excitement to get a free ride as the trolley glided downhill. Next, Jack learned to perform Jumper's job as signalman by waiting patiently with Jumper in the signalman's hut and listening for the number of blasts from the approaching locomotive drivers. Each track was assigned a different number. If the driver gave one, two, or three blasts, Jack switched the signals in the appropriate manner, altering the direction of travel so that oncoming trains would not collide. If the driver gave four blasts, Jack collected the key to the coal shed and carried it out to the driver. His performance was so unerringly correct that he earned the name "Jack the Signalman" (Fig. 2).

Figure 2. Jack the Signalman with his coworker, Jumper Wide. Photo courtesy of Paul Screeton.

On one occasion, a prominent lady traveling from Cape Town to Port Elizabeth saw to her horror that the signals in the train yard were being changed by a baboon. When executives in Cape Town received her indignant report, their first reaction was disbelief. When she insisted that her account was true, they sent a delegation of inspectors to Uitenhage. Jumper and Jack were dismissed from duty. But once again Jumper persuaded the inspectors that he (and Jack) could do the job. He challenged the inspectors to give Jack a rigorous test of his skills, and Jack passed with flying colors. He even looked in both directions each time a signal was changed, apparently checking to make sure that trains passing in the yard would be on different tracks. From that day on, Jack the Signalman received daily rations and was given an official employment number. After a long and successful career, he died of tuberculosis in 1890 (du Plessis, n.d.).

Accounts of Jack the Signalman and other working baboons spread widely and helped to convince 19th-century South African farmers that baboons were intelligent creatures. But stories like Jack's revealed little about baboons' social skills. Other anecdotes were more illuminating.

Traveling through what is now Namibia in 1836, the British explorer James Alexander (1838) reported that the Namaqua people sometimes kept baboons as goatherds. These baboons reliably followed the herd of goats during the day, keeping the animals together and giving alarm calls if a predator was spotted. At dusk, they guided the flock back to the compound, sometimes riding on the back of the largest goat.

The Namaquas' domestication of baboons as goatherds was apparently passed on to the local European farmers, because it persisted until very recently. In 1961, the German naturalist Walter Hoesch described the behavior of an adult female, Ahla, who served as a faithful tender of livestock on a Namibian farm for over seven years (Hoesch 1961). She was the third baboon that the owners had employed as a goatherd. As part of her work, Ahla groomed the goats and led them back and forth to the fields every day, again often riding on one of the larger goats (Fig. 3). Like the baboons who served as shepherds for the Namaqua, Ahla gave alarm calls when predators were spotted. She also recognized immediately when a goat or kid was missing from the herd and searched anxiously for it, giving what Hoesch describes as "ho ho" calls until she found the missing animal.

Most intriguing for our purposes was Ahla's ability to recognize the kinship relations that existed among individual goats. In the evenings, the kids were often separated from their mothers in a different barn.

Figure 3. Ahla the goat-herding baboon grooms one of her charges.

Figure 4. Ahla carries a kid to its mother.

When this happened, Ahla became "not only eager but really a maniac" in her efforts to reunite kids with their mothers (Fig. 4). Hoesch reported:

When Ahla comes home in the evening after feeding, she will first go to the enclosure and from there through a door to the lambs' enclosure. From here, she can only hear the adult animals but not see them. Once she hears from inside the voice of a lamb that is calling for its mother, she will retrieve the correct lamb and jump through the opening between the two enclosures and put it underneath the mother so it can drink. She does this flawlessly even when several other mothers are calling and several lambs are responding at the same time. It seems impossible that she does this solely on the smell of the animals. She also retrieves lambs and brings them back even before mother and infant have begun calling. Apparently, she knows every animal in the herd but it seems unclear how she effectively recognizes them. Mrs. Aston [the farm's owner] noted that "No local personnel and also no white person would be able to assign correctly the 20 or more identically looking lambs to the mothers. However Ahla is never wrong." (Hoesch 1961:299)

Ahla's determination to ensure that each kid remained with its mother sometimes interfered with her work as the farmer's assistant. When a ewe gave birth to twins, the farmer often tried to foster one of the kids onto a ewe that had lost her own offspring. Ahla would have none of this. She *knew* which kid belonged with which mother, and at every opportunity she doggedly returned the fostered twin to its real parent. Hoesch noted that Ahla was never trained to recognize the kinship relations between mother goats and their kids: "she does things that she has never observed and that she has never been told." Occasionally, though, "she was punished with a belt when for instance she took a lamb up into the top of the trees where she slept."

Intrigued by Ahla's tale—but at the risk of losing what remained of our scientific dignity—we traveled to Namibia in 2006 in search of a goat-herding baboon. Unfortunately, the practice appears to have fallen away, although through our colleague Conrad Brain we did speak with several people who had owned, or knew someone who had owned, a herding baboon. One farmer, Walter Utz, reported that he had employed a female baboon, Bobejaan (the Afrikaans word for baboon), as a goatherd well into the 1980s at his farm near Otjiwarongo. Bobejaan led the goats to and from their pastures each day, watched for predators, groomed the goats, and kept them free of ticks. She also groomed Mr. Utz's two Simmentaler bulls, which would wait every evening by the gate for Bobejaan to return from the fields with the goats. Mr. Utz remarked that Bobejaan could also recognize each goat's offspring by both voice and sight, and that she had learned to do so without any training. Bobejaan became so effective at guarding her herd that she was eventually trapped and killed by local rustlers who wanted to steal the goats. Mr. Utz concluded the conversation by stating that he would certainly employ another baboon if he ever again decided to keep goats.

As we will discuss later in this book, baboons and other monkeys may be unique among animals in recognizing the close bonds that exist among other members of their group. The accounts of goat-herding baboons, though anecdotal, suggest that this ability also allows baboons to recognize the relationships that exist among members of another species. They suggest, too, that the first people to recognize baboons' unusual social skills were not European or American scientists in the late 20th century, but the Namaqua people who kept baboons as goat-herds many hundreds of years ago.

THREE

Habitat, Infanticide, and Predation

It was not long before we came to realize that the life of the baboon is in fact one continual nightmare of anxiety.

EUGENE MARAIS, 1939: *MY FRIENDS THE BABOONS*

The baboons' habitat

The Okavango Delta in northwestern Botswana has been justifiably described as one of the most spectacular environments in the world. It is also unusual among African habitats for its striking changes from one season to another. Between October and May—summer in the southern hemisphere—the delta is a classic African savannah-woodland. Vast open plains covered with golden, knee-high grass are broken by patches of woodland that range in size from a few hundred square meters to fifty hectares or more. Seen from the air, the patches of woods look like lush green islands surrounded by a golden sea. The savannah and the woodlands not only look different, they have a very different feel. As you walk across the savannah at midday, the heat of the sun is intense. Temperatures regularly rise above 40° Celsius (104° Fahrenheit). The grass shimmers in the heat, and without a hat your head soon begins to swim. But once you enter a patch of forest you come under the shade of enormous trees, some of them 25 meters (80 feet) high. The harsh light dims and the temperature drops. Even on the hottest day, walking in the woodlands of the delta is relatively pleasant.

October through March is also the rainy season in Botswana, but this does not necessarily mean there will be rain. The delta's annual rainfall is around 400 millimeters (16 inches), much of it coming during a few intense storms in December, January, or February. The variation in rainfall from one year to the next, however, is huge. In many years, the rains are miserly and desultory, prompting local residents to shake their heads and mutter about the worst rainy season in decades. In other years, the rain is relentless. The grass grows furiously, and both the baboons and other animals disappear in a towering mass of green. This can prove awkward when the grass conceals a buffalo, lion, or disgruntled baboon.

While the delta bakes in the summer heat, rain in Angola swells the Cuito and Cubango Rivers, which join together, flowing southeast, to form the Okavango. As its waters rise, the Okavango flows first through a small corner of Namibia and then into Botswana, arriving at the border post of Mohembo sometime in December or January. Roughly 80 kilometers (50 miles) southeast of the border, the Okavango breaks into several tributaries, each of which breaks into several more. The flood waters then begin their leisurely journey through the swamplands, traveling more than 450 kilometers (280 miles) during the next seven months before reaching the town of Maun some time in the midwinter months of June and July. The water spreads and expands to create a large, triangular delta that extends for hundreds of square kilometers. Streams fill their beds and spill over their banks, flooding the savannah. Grasslands are immersed in water that ranges anywhere from several centimeters to two or three meters deep. Patches of woodland are now no longer metaphorically but literally islands, completely surrounded by water. Roads become impassable, camps are isolated, and the only means of travel is by plane, boat, dugout canoes (mekoros), or wading through the shallower crossings (Mendelsohn and el Obeid 2004).

Charles John Andersson, one of the first Western explorers to visit the delta in 1856, marveled at both its beauty and the apparent miracle that such a volume of water could be concealed in the otherwise hostile Kalahari Desert. "On every side, as far as the eye could reach, lay stretched a fresh sea of water, in many places concealed from sight by a covering of reeds and rushes of every shade and hue, while numerous islands, spread over its surface, and adorned with rich vegetation, gave to the whole an indescribably beautiful appearance" (Andersson 1856:474) (Fig. 5). Local people had told Andersson's contemporary, David Living-

Figure 5. Camp Island at the height of the annual flood. Photograph by Anne Engh.

stone, of "a country full of rivers—so many that no one can tell their numbers—and full of large trees" (Livingstone 1858:75), but Andersson and Livingstone found this difficult to believe.

Like the summer rains, the size of the flood is highly variable. In good years, the flooding can cover over 16,000 square kilometers (6,000 square miles); in poor years the swamps shrink to less than half this size. Since the early 1990s, most floods have been significantly smaller than in the previous 50 years, probably due in part to increasing diversion of water from the Okavango River in Angola and Namibia. Proposed pipelines and dams in these countries pose a serious threat to the Okavango Delta, but that worrying prospect is a subject for another time.

The magnitude of the flood is so important to local people—for their livestock, their drinking supply, their subsistence farms, and the wildlife on which their incomes depend—that the flood is watched, measured, and reported with the kind of obsessive attention usually reserved for the last few weeks of pregnancy or the World Cup standings. As the floodwaters enter Botswana, government officials measure the Okavango's flow in cubic millimeters per second, enter the results on a graph that compares the flow, week by week, with legendary floods of previous years, and distribute the graph throughout northwestern

Botswana, where it is posted in bars, restaurants, and shops. No one talks about anything else. In a good year, the flood arrives at our research site in April or May and does not recede until October or November, when, with luck, the rains begin again. A bad flood compounded by poor rains, on the other hand, means dry wells, dying livestock, and the grim heat and dust of October, known locally as "the suicide month."

The Okavango and its annual flood form the backdrop for our study of baboon society, just as the ecology of early 19th-century England formed the backdrop for Jane Austen's studies of life in the English counties. And just as the comforts of the landed gentry separated Austen's characters from the harsh poverty found elsewhere in England, so the richness of the delta and the timing of its annual flood buffer our subjects from the harsh seasonality that confronts baboons elsewhere on the continent. During the summer, trees in the Okavango provide an extraordinary bounty. Among other trees, two species of figs (*Ficus* spp.), three species of acacia (*Acacia* spp.), the real fan palm (*Hyphaene ventricosa*), African mangosteens (*Garcinia livingstonei*), and marula trees (*Sclerocarya birrea*) occur throughout the baboons' range. The figs are particularly important because they repeat their fruiting cycle throughout the year, providing one lush harvest after another. On a typical summer's day, it is not unusual for the baboons to feed all morning around a single huge sycamore fig (*F. sycamorus*), and then spend the afternoon lounging about, dozing, grooming, and playing. When summer's richness comes to an end the flood arrives, bringing unlimited water, new grass, the ripening fruit of the sausage tree *(Kigelia africana)*, and the extraordinary bounty of jackalberry trees (*Diospyros mespiliformis*), huge members of the ebony family that conveniently bear their fruit in the middle of the winter. Eighty baboons can feed in a large jackalberry for three or four days without completely depleting it. As a result of its ecological richness, the Okavango Delta boasts the highest reported density of baboons in all of Africa (Hamilton et al. 1976).

The Okavango Delta is one of world's most spectacular places, but conducting research there is not without its perils or inconveniences. Our site is remote, and in an average flood year we can drive to Maun, the nearest town, during only a few months. When we first arrived in July 1992, we looked forward to driving to Maun after the floodwaters had receded, because we imagined the trip to be an extended game drive with countless possibilities for observing rare species of birds and mammals. The first journey disabused us of our naïveté. The drive to Maun is instead a disagreeable ordeal with endless possibilities for becoming lost, getting stuck, overheating the engine, puncturing tires, and being chased by elephants. On a good day, the 90-kilometer journey takes between four and five hours. The "road" is in fact a deeply rutted track through glutinous sand and mud, and it is not unusual to have to dig the car out of deep pits several times during a trip. Because there are no mountains or hills in the Okavango, and because tracks often circle back on themselves to avoid fallen trees or sand pits, it is easy to become disoriented without a GPS (global positioning system) monitor.

When the flood arrives, a vehicle becomes little more than a large lawn ornament, stranded on an island. Although it is undeniably romantic to live in a remote campsite completely surrounded by water, supplies can be a logistical nightmare (colleagues who work in far more remote areas than ours will rightly scoff at these complaints). Now the only access from Maun is by boat or plane to a nearby lodge, roughly half an hour downstream. Supplies are ferried from the lodge to camp in a small 25-horsepower aluminum boat, optimistically guaranteed by the manufacturer against attack by crocodiles but not by hippos. The trip is literally exhilarating, because there is always a good chance of a hippopotamus encounter.

It is no accident that, in their accounts of the Okavango Delta, David Livingstone and Charles Andersson devoted more pages to narrow escapes with hippos than with any other animal. Hippos appear to have only one emotional state—anger—and no faculty of reason. Each year, they kill more people in Africa than any other mammal. Our research assistants' mekoro has been charged, overturned, or bitten by a hippo on numerous occasions. Everyone in the Okavango has experienced several harrowing, narrow escapes from hippos, and in the process developed a profound and enduring hippo phobia. There is little to recommend the hippo.

In part because of the flood, we conduct all of our research on foot, wading with the baboons from island to island. Following baboons on foot for hours at a time can be unremittingly tedious, but the tedium is occasionally punctuated by moments of intense fear. As a result, anyone who has ever worked at the research site can, without the smallest provocation, transmogrify seamlessly into a swashbuckling pedant, recounting numerous, harrowing tales of buffalo they have escaped, elephants they have evaded, lions they have stumbled upon, snakes they have avoided, and so on. The weary listener cannot even shift the topic of conversation by asking about camp life, because the eager raconteur will seize this opportunity to describe how he or she almost walked into an elephant when leaving the shower, or was stalked by lions when going to her tent at night. These encounters not only enliven cocktail party conversation (at least for the narrator) but also serve as sobering reminders that the challenges facing the baboons are real—not artifacts of human training or captivity—and that the solutions they come up with are the result of strong selective pressure.

Infanticide

From the time it was discovered by Western explorers, the Okavango Delta has been likened to the Garden of Eden. From the cool comfort of a wooded island, Andersson described a wealth of trees never before encountered, "many yielding an abundance of palatable and nourishing fruit. ... The arboreal scenery, indeed, in some places exceeded in beauty any thing that I had ever seen." And the "animal life was almost on a par with the exuberant vegetation. Rhinoceroses, hippopotami, buffaloes, sassabys, hartebeests, pallahs, reed-bucks, leches, &c. were constantly seen, and every day some game animal or other was shot" (Andersson 1856:460–461).

Andersson's allusion to Eden is certainly apt, but for the baboons the garden is not without its serpents. Some are of their own making. Because the Okavango Delta contains one of the highest densities of baboons in Africa, each group's range covers an area of only 3–4 square kilometers, considerably smaller than that found among baboons elsewhere (Hamilton et al. 1976). The high density of baboons seems to create intense competition among males for access to mates. This, in turn, leads to high rates of turnover in the male dominance hierarchy: an alpha male can expect to retain his status for only seven or eight months. Perhaps as a result, and as we discuss in Chapter 4, newly risen alpha males go to great lengths to maximize their reproductive success during the brief time when they hold the alpha position, even if it means killing infants fathered by the male they have just defeated. Between 1992 and 2006, confirmed or suspected infanticide caused the majority of infant mortality, accounting for at least 53% of all infant deaths (see also Cheney et al. 2004). By comparison, male infanticide occurs relatively infrequently among baboons in East Africa, where male reproductive skew is less extreme and both baboon densities and turnover rates among males are lower (Alberts et al. 2003; Palombit et al. 2000).

Predation

Other serpents in the garden are beyond the baboons' control. Herds of ungulates attract predators, and the abundance of fish supports one of Africa's largest populations of crocodiles (*Crocodilus niloticus*). Predation accounts for the vast majority of deaths among male, female, and juvenile baboons. Between 1992 and 2005, lions (*Panthera leo*) and leopards (*Panthera pardus*) were known to kill 19 baboons in our group. In 16 other cases, predation was highly suspected, either because we saw the tracks of leopards or lions near the site of a baboon's disappearance, heard alarm calls, or found fur, bones, or blood (Cheney et al. 2004). Many more adults and juveniles simply disappeared, having been seen, apparently healthy, the day before. We suspect that adults and juveniles who forage on the periphery of the group or lag behind during water crossings are often picked off by leopards, lions, or spotted hyenas (*Crocuta crocuta*). We estimate that predation has accounted for roughly 95% of all adult female deaths. In 2002, 25% of the group's 28 adult females disappeared. Eighteen months later, between September 2003 and August 2004, another 26% of the adult females died of confirmed or suspected predation. Bulger and Hamilton (1987) report a similarly high proportion of deaths among adult females due to predation in 1984 (Fig. 6).

Predation is especially common during the flood, when baboons, buffalo (*Syncerus caffer*), giraffe (*Giraffa camelopardalis*), zebra (*Equus*

Figure 6. Lions are one of the baboons' primary predators. Photograph by Chris Harvey.

burchelli), kudu (*Tragelaphus strepsiceros*), and other ungulates are forced to make long water crossings from one island to another. Because they are vulnerable to crocodiles, lions, and other predators during these crossings, the animals quickly learn the shortest and shallowest routes, moving from island to island by predictable paths. Inevitably, lions and other predators also learn where to conceal themselves and wait. We have seen lions kill baboons during water crossings on several occasions.

Water crossings, as a result, are fraught with anxiety. Long before they enter the water, the baboons sit at the island's edge, nervously grunting and looking out toward the island they hope to reach. Any movement on the water's surface elicits a chorus of alarm calls and brief flight. Once they seem satisfied that the coast is clear, adults begin to cross. Reluctantly, the juveniles follow, some grunting nervously, others moaning or screaming, and others running to leap on their mothers' backs, anxious to get a ride (Fig. 7). Many of the baboons adopt their own idiosyncratic way of crossing, seemingly desperate to keep particular parts of their bodies as dry as possible. Some walk, trot, or jump on their hind legs, holding both arms high in the air, while others seem particularly concerned about their left hand, or the tip of their tail (Fig. 8). If the water is deep, juveniles and even adults may be forced to swim. The whole spectacle is chaotic and amusing to the human

Figure 7. Balo carries her young daughter Domino through a water crossing. Photograph by Keena Seyfarth.

Figure 8. Baboons adopt a variety of idiosyncratic water-crossing styles. Lissa has perfected the bipedal hop. Photograph by Dawn Kitchen.

observer but deeply distressing for the baboons, who are out of their element and vulnerable to any predator that lurks in the water or along the too-well-traveled path. The baboons' anxiety is probably heightened by the knowledge that lion ambushes are unpredictable, uncontrollable, and likely to result in the simultaneous deaths of many individuals.

On September 9, 2003, just before noon, our colleague Anne Engh was following the baboons as they foraged at the edge of two wooded islands not far from camp. The group was widely dispersed. Most of the males and females had crossed the drying floodplain and begun digging for lily roots in the shallow water near the river. Many of the juveniles remained in the woods, feeding and playing among the trees. Anne was looking for Jeanette, an old adult female who was the next animal on her list to be sampled. She was a bit concerned because she had not seen Jeanette all day. As she headed out onto the plain, Anne heard alarm calls from the males near the river and saw them running in all directions. This was a bit worrying, but baboons are always nervous near the river and they had been giving sporadic alarm calls all morning, so Anne continued to look for Jeanette. But as she walked away from the woods and further out onto the plain she became increasingly puzzled: no sign of Jeanette anywhere, and in fact no sign of any baboons at all. She was utterly alone. Then she froze in her tracks. A few meters away was a lioness, crouched in the grass and staring right at her. Two thoughts pounded in Anne's head: first, *never run from a lion*. A lion is like a big kitten, and if you run you will be its ball of yarn. Second, *look out for more trouble*. Female lions are part of a kin-based pride and lions hunt in groups. Slowly and deliberately, Anne walked back toward the woods and the safety of climbable trees.

By the time she reached the woods all hell had broken loose. The males near the river, females on the floodplain, and youngsters in the woods had scattered and were now widely separated into two groups, each on its own island. Everyone was high in a tree, and the air was filled with loud alarm calls. And while the trees were filled with baboons, the only animals Anne saw on the ground were lions—in the woods, on the floodplain, everywhere.

The mass attack that day involved six lions. Seven baboons were killed: adult females Jeanette, Heloise, Bennet, and Capricorn, and juvenile males Neptune, Molomo, and Lefsa. Jeanette's death left as orphans her adult daughter, two juvenile sons, and a 10-month-old infant. Heloise's death left her 18-month-old son.

After the attack, the group remained separated for two days. On the second day, animals in one subgroup began giving contact barks, the loud call that baboons give when they become separated from their companions (we discuss this intriguing vocalization in Chapters 8 and 9). Animals in the other subgroup heard them and moved cautiously to rejoin. Each group approached the other along the edge of their island, moving from tree to tree, never venturing onto the ground. Every hundred meters or so they stopped to stare vigilantly around them, nervously giving alarm calls. The reunion took place several hours later.

Five months later, on February 16, 2004, it happened again. This time Anne and her assistants, Rebekah Hoffmeier and Alec Mokopi, had lagged behind the group to take a GPS reading. Suddenly, about 50 meters ahead, the baboons began screaming, alarm-calling, and running in every direction. Swirling around them were three lionesses, one with a baboon in her mouth (Fig. 9). As Anne wrote in her blog, "With heavy hearts, we watched the lioness disappear into the bushes with one of our troop members. It was getting late, and we knew that our chances of finding all of the remaining baboons were slim that afternoon, but we really wanted to know who had been killed, so we started to make a list of who was left. Just as we began, we heard horrible rumblings and bone-crunching coming from the bushes behind us. Out popped a pair of lionesses wrestling and growling over the carcass of a big baboon."

Figure 9. Successful predator attacks are especially stressful to the victims' close relatives. Photograph by Chris Harvey.

The attack of February 16 involved three or four lions and killed two adult baboons, Cyrano, a male, and Sierra, a female. Sierra was dear to our hearts: a daughter of Sylvia, niece of Selo, and a central member of the highest-ranking matriline. Her death left as an orphan 18-month-old Margaret, who survived and retained her high rank, supported in grooming and alliances by her grandmother, aunts, and cousins.

Baboons can defend themselves against lions only by being vigilant, giving alarm calls, and fleeing into the trees. If they are lucky enough to see lions before they attack, or if they come upon lions that are not hunting, the baboons' immediate response is to climb trees and give alarm calls. Then, more subtly, over the next hour or two, they change their direction of travel by approximately 180 degrees (Kitchen et al. 2004). We behave the same way. Lions usually move off at our approach, but on more than one occasion we have found ourselves up a tree, sitting on a limb alongside a row of several baboons, all of us looking down at the lions below.

Because they sleep in trees, baboons are generally safe from lions at night. Relations with leopards are more complex. At night, leopards have the advantage. They attack baboons by climbing into their sleeping trees, sometimes returning several times on the same night to take yet another baboon (Busse 1982). It is often possible to reconstruct a leopard kill on the morning after. On several occasions when the baboons have slept near our camp, we have heard screaming and alarm-calling during the night. The next morning, we have not only discovered that a baboon is missing but also found leopard tracks, drag marks, baboon fur, and even, sometimes, the carcass itself.

Leopards also hunt baboons by hiding in bushes near their sleeping trees and pouncing on the first animal to descend at dawn. One morning, for example, we heard a chorus of alarm calls emanating from a sleeping site near camp. Racing out to the sleeping site, we found a huge male leopard on a termite mound, surrounded by an angry circle of lunging, screaming, and alarm-calling baboons. After several minutes, the leopard spied a break in the baboon circle and raced to his escape. We searched the bushes and found the body of Nanook, an adult female. The leopard had cached Nanook's carcass, expecting to return to it after the baboons had left the area. Nanook left an infant son who died soon afterwards, despite the ministrations of Nanook's adult male friend, Wanda.

The baboons' assault on Nanook's killer demonstrates how, during daylight, the balance of power between leopards and baboons can shift in the baboons' favor. As they forage through the woods and savannah, baboons cut a wide swathe and make a fair amount of noise. A leopard resting in a shady bush or tree probably hears them coming and hides or runs away long before they arrive. But if the baboons are able to isolate a leopard in a bush, tree, or aardvark hole, they immediately surround it, screaming, alarm-calling, and lunging at it, seemingly without fear. Although male baboons, with their size and enormous canines, are much better equipped than females to fight a leopard, the mass mobbing involves baboons of every age and sex. Juveniles, adult females, even mothers with young infants join to form a huge, hostile mob that tries to corner the leopard. The attack continues even after some baboons have received slashes on their arms, legs, and face that open up huge wounds. One old, low-ranking female, Martha, had a particular antipathy toward leopards. She was always in the vanguard of mobbing attacks. Over the years, she recovered from several leopard-inflicted injuries before finally being killed at the age of 20. During one attack, the leopard was guarding two tiny cubs in the hollow of a sausage fruit tree. For over half an hour, the baboons lunged at the hollow as the leopard swiped back. Although we quietly sided with the baboons during most predator encounters, we loathed the prospect of seeing the cubs ripped to shreds and were relieved when the baboons finally moved away. Martha was the last to leave.

These mob attacks are not just displays of bravado. We have seen baboons injure several leopards, and they are known to have killed leopards in other areas of Africa (Cowlishaw 1994). Rangers in the Kruger National Park of South Africa describe finding a dead leopard, covered with bites, lying in the grass next to two dead baboons. Not surprisingly, when baboons begin mobbing, the leopard tries to break free by finding a gap in the circle of baboons. During one such escape, this gap was made up of our colleagues Dawn Kitchen, Jim Nicolson, and Press Kehaletse. As the leopard raced by, it knocked Press over and began to scratch and bite him. Within seconds, a horde of more than 20 baboons leapt onto the leopard and drove it away. Several females were involved in this counterattack, including one carrying a young infant. Luckily, Press escaped with only superficial wounds.

In contrast to lions and leopards, hyenas and wild dogs (*Lycaon pictus*) typically elicit only scattered alarm barks from the baboons. This is somewhat surprising, because hyenas do prey on baboons, especially when they are hunting in a group. Anne Engh observed a hyena calmly grab and kill a baboon that was foraging near it. The hyena then slowly walked away with the carcass to scattered alarm barks, followed by its two companions. We strongly suspect that hyenas opportunistically pick off baboons that are foraging on the periphery of the group, because we have often stumbled upon a lurking hyena while searching for an animal to sample. Although hyenas are much larger than baboons, the baboons will chase a hyena if they encounter it alone. Our colleagues Thore Bergman and Jacinta Beehner once saw several adult male baboons drop from a tree to attack a solitary hyena that was passing underneath.

Finally, crocodiles occasionally attack baboons when they are feeding in floodplains or foraging next to the river. On three occasions we have seen a crocodile leap from shallow water and attack an adult male who was foraging or drinking nearby. In two cases, the male jumped away without being bitten. In a third attack, however, a relatively small (2 meter long) crocodile grabbed the group's alpha male and bit him on the face, arms, and leg before he could escape.

Although crocodiles may not account for as many baboon deaths as lions and leopards, anyone watching baboons near water would have no doubt that they fear and despise these reptiles. As we have already mentioned, baboons become extremely vigilant before crossing water, giving alarm calls to any disturbance. When they forage near the river, they are even more anxious, and their nervousness often leads young animals to give alarm calls to partially submerged hippos, diving ducks, and even floating elephant dung. The slightest boil on the water's surface elicits a chorus of crocodile alarms from the baboons and a race away from the water's edge.

Predation and stress

When Marais described the life of a baboon as "one continual nightmare of anxiety," he was referring to baboons' fear of predators. His intuition rings true. Because predator attacks are unpredictable, uncontrollable, and unremitting, and because witnessing an attack is traumatic in itself (as we can attest), the threat of predation could potentially result in chronic anxiety.

When a baboon, human, or any other mammal faces an acute emergency—like being chased by a lion—the body mounts a stress response. Minutes after the emergency is detected, the adrenal cortex begins to pump glucocorticoids into the blood. The elevated levels of glucocorticoids act both to mobilize energy reserves and to curtail nonessential metabolic activity (Sapolsky 2002, 2004). While adaptive over the short term, this flight-or-fight response is physiologically costly if it persists. Occasional stressors can be tolerated, but long-term, chronic stress drains the body of essential reserves and has many other deleterious consequences. Some hallmarks of prolonged stress are atherosclerosis, cardiovascular disease, and a compromised immune system.

Predation is a pervasive and inescapable part of baboon life. It is the single most important cause of adult mortality, and it affects individuals of all dominance ranks equally. Even the survivors suffer: in months when a group member is killed by a predator, glucocorticoid levels in males and females are significantly higher than at other times. Attacks that result in group separation—like the lion attack that divided the group for two days—produce significantly greater stress than less intense encounters (Engh et al. 2006b). Predation is especially stressful for females whose close relatives are killed. Females who lose a close relative to predation have significantly higher glucocorticoid levels during the following month than do females who experience no such loss. Female baboons, in other words, experience what in humans we would call bereavement. And, as we will describe in Chapter 5, in the subsequent weeks and months females respond to their grief in a very human way: by seeking out others and enlarging their social network (Engh et al. 2006b).

Predation and infanticide are the dark side of the Okavango Delta. Along with the open savannah, lush islands, huge fruiting trees, and the annual flood, they set the stage for our study of baboon society and the metaphysics that underlies it.

FOUR

Males: Competition, Infanticide, and Friendship

The generality of the male sex cannot yet tolerate the idea of living with an equal...It is only a man here and there who has any tolerable knowledge of the character even of the women of his own family.

JOHN STUART MILL, 1869: *THE SUBJECTION OF WOMEN*

Baboons are a politically incorrect species. Males are much bigger than females and are best described as bullying hooligans. Although they sometimes form close, enduring friendships with females, they remain dominant to them and often express their affection through threats and violent chases. Unlike females in less sexually dimorphic species like vervets and macaques, female baboons almost never form alliances against a male, except when the male is attacking their infants. Nevertheless, females form the core of baboon society. They remain in their natal groups throughout their lives, while most males emigrate to neighboring groups as young adults. Moreover, while females live comparatively tranquil, subtle, and complex lives—and often survive to more than 20 years of age—the lives of most males are nasty, brutish, and short. Because the strategies associated with success in male baboon society are relatively simple and uncomplicated, we discuss males first.

An infant baboon is born after six months' gestation and remains dependent on its mother for roughly one year (we therefore define infants as those 12 months of age or younger). The youngest infant we have known to survive its mother's death was nine months old when her mother died.

By about 15 months of age, young juveniles are feeding and moving independently (Altmann 1980). Not long afterwards, male and female baboons begin strikingly different life trajectories. By five years of age, males are considerably larger than females. By the time they become fully adult—females at around six and males at around nine years of age—males weigh twice as much as females. They also develop large, dog-like muzzles and enormous canine teeth.

At some point after the age of nine, most males leave their natal group and join another. The timing and causes of a male's departure are not always clear. Some simply leave, unpredictably, before they have shown any inclination to challenge other males. Others rise in rank through the male dominance hierarchy—sometimes attaining the alpha position—and leave only after they have been defeated. A few never leave.

The number of adult males in a baboon group at any given time ranges widely, from as few as 3 to as many as 12. Regardless of their number, however, the males invariably form a linear, transitive dominance hierarchy based on the outcome of aggressive interactions (a linear, transitive hierarchy is one in which individuals A, B, C, and D can be arranged in a linear order with no reversals that violate the rule 'if A dominates B and B dominates C, then A dominates C'). Although the male dominance hierarchy is linear, transitive, and unambiguous over short periods of time, rank changes occur often (Kitchen et al. 2003b), and a male's tenure in the alpha position seldom lasts for more than a year.

Displays, fighting, rank, and sex

When an immigrant arrives, he challenges the males in his new group with aggressive displays. Male chases and displays are a daily occurrence; physical fights, by contrast, are relatively rare. This is not unusual. In most animal species, competitive interactions typically take the form of vocal, visual, or gestural displays—loud calls, threatening postures, and other behaviors that allow opponents to assess each other's fighting ability without always escalating to an actual attack.

Regardless of the species involved, displays have one essential property: they involve signals of competitive ability that are difficult, if not impossible, to fake. Male red deer (*Cervus elaphus*), for example, challenge and assess one another by the rate, amplitude, and pitch of their roars (Clutton-Brock and Albon 1979). These acoustic features are reliable indicators of size and endurance because only large males in excellent condition can produce loud, low-pitched roars at high rates. In much the same way, songbirds challenge and assess one another by the size of their song repertoires. Males with large and complex song repertoires are often older, more experienced, or in better condition than others. Repertoire size is therefore a good indicator of a male's age and condition, which in turn is correlated with his competitive ability (Searcy and Nowicki 2005; for another review see Andersson 1994).

Because natural selection favors the skeptical observer and acts against any individual who allows himself to be duped by traits unrelated to fighting ability, the only displays that persist over time are "honest" indicators of a male's condition. The displays are honest because they are too costly for males in poor condition to produce and maintain (see Andersson 1994, Vehrencamp 2000 for reviews). Displays will always be more common than actual fighting because, regardless of the competitive ability he brings to the table, it invariably pays for a male to display first, before the fight escalates and results in injury (Fig. 10). Avoiding injury is of paramount important because the cost of being injured almost always outweighs the benefits to be gained from any single dispute (Maynard Smith 1982; see Bradbury and Vehrencamp 1998 for review).

Male baboons' competitive displays take the form of violent chases and loud "wahoo" calls. Wahoos satisfy all of the criteria for a classic competitive display. They are extremely loud, low-pitched calls that can be produced only by large, fully adult males. They are costly to produce not just because of their loudness and low pitch but also because males give them in long bouts, often as they race through the group or bounce through trees, leaping from branch to branch. A wahoo display is therefore an exhausting demonstration of a male's stamina and coordination. And lest there be any doubt that wahoo displays are surrogates for actual fighting, male baboons often engage in wahoo "contests": one male's wahoos elicit challenges from two, three, four, or many other males, all of whom respond by leaping through the trees and wahooing themselves. The contest continues until, one by one, the males drop out, exhausted. Usually, only one of the most dominant males is left calling at the end.

Wahoos really do seem to provide a reliable indicator of a male's competitive ability. As research by our colleagues Dawn Kitchen and Julia Fischer has shown, high-ranking males are more likely than low-

Figure 10. Male baboons assess each other's competitive abilities in chases, wahoo contests, and canine displays. Betelgeuse yawns to signal his willingness to defend his estrous female, Champagne. Photograph by Anne Engh.

ranking males to enter wahoo contests. They also give wahoos at the highest rate and produce wahoos with longer and louder *hoo* syllables than the wahoos of other males. As males age and fall in rank, they are less likely to enter wahoo contests—and with good reason, because their wahoos become less and less impressive. While the wahoos of a young, vigorous, high-ranking male are thunderously loud, intimidating calls, the wahoos of old males are weak, hoarse, *waagh*s with little *hoo* left at the end (Kitchen et al. 2003b; Fischer et al. 2004a).

The predictable relation between a male's rank and the quality of his wahoos allows competitors to assess each other without actual fighting. Males of very disparate ranks seldom engage in wahoo contests,

Figure 11. Some displays between males escalate to physical fights. Photograph by Adrian Bailey/Aurora.

presumably because subordinate males can assess, through their rival's wahoos and behavior, that they are outmatched and that discretion is the better part of valor. Escalated fights between males of very different ranks are therefore rare, occurring only when the contested resource is extremely valuable: meat, a sexually receptive female, or an infant threatened by infanticide (Fig. 11). By contrast, wahoo contests involving males of similar rank—whose wahoos sound more alike—occur at high rates. They are longer, occur for unpredictable reasons, and are more likely to escalate to physical fights and wounding (Kitchen et al. 2005a).

At any one time, a baboon group is likely to contain several males at different stages in their life history. Some are young adult natal males of nine or ten years who have yet to disperse. They have impressive, fully erupted canines, and their testosterone levels are high (Beehner et al. 2006). They challenge other males with wahoo displays and chases, by yawning to display their undamaged canines, and by harassing females. If his rival is not intimidated, a young, testosterone-charged male may eventually escalate his challenge into an actual fight. He is making his move up the hierarchy.

Other males of the same age or a little older are immigrants from another group. They, too, are in prime condition, with sharp, undamaged canines and impressive wahoos. One of these individuals is typically the group's alpha male. Some alpha males have high testosterone and aggressively assert their dominance by initiating wahoo contests or walking assertively directly at their rivals. Others have surprisingly low testosterone levels and seem to maintain their status effortlessly, by sheer force of personality (Alberts et al. 1992; Sapolsky 1992, 1993, 2005; Beehner et al. 2006).

Invariably, the group also includes several males who have recently fallen in rank or never attained the alpha position but who are still relatively young and in good condition. Some of these males are protecting infants they may have fathered, others appear to be biding their time.

Finally, there are older males with worn, chipped, badly stained canines and hoarse, exhausted wahoos that are unlikely to impress anyone. As in so many cultures, these old fellows are a mixed bag. Some are once-great individuals whose loss of rank appears to have sent them into a tailspin of depression. Their testosterone levels plummet and they seem to age before our eyes. Most die within a few years of their fall, often before they reach 15 years. Others are down but by no means out. They continue to defend infants and challenge other males. In 2000 and 2001, a tenacious male known as Third Man moved up and down the hierarchy, from top to bottom and back, occupying seven different rank positions over a 12-month period. When he was up he wahooed impressively, when he was down he did not. Still other senior males have teeth that are worn to nubs or entirely gone. Although they move slowly and seldom engage in wahoo displays, they remain sufficiently ornery and resolute to seize the opportunity for a surreptitious mating, to defend a favorite juvenile, or to kill an impala or vervet. One such male, Wanda, lived beyond age 20, the only male in the group ever known to achieve this actuarial milestone.

Interestingly, once a male has reached adult size there seems little relation between his size and either the quality of his wahoo displays or his rank (Kitchen et al. 2003b). Craft, guile, and persistence may be more important. Although a very small male might never attain alpha status, it is also the case that not all big males do so. Because a male's tenure in the alpha position typically lasts for less than a year, there are always males of roughly similar size in the group who formerly occupied this position, or who will shortly attain it. A male's fighting ability determines his rank, but fighting ability appears to be related more to skill and ambition than to size. Indeed, as we discuss below, some alpha males appear almost to abdicate their position once they have fathered infants that may require protection from immigrants.

Competition for high rank is the central feature of a male baboon's life because rank translates directly into reproductive success. The alpha male accounts for the vast majority of matings, particularly those that occur around the time of ovulation (Bulger 1993; Kitchen et al. 2003b). Female baboons experience a reproductive cycle of approximately 28 days. Roughly 10 days before ovulation, the skin around their perineum turns pink and begins to swell. During this time, females solicit matings from males, but only juveniles and adolescents usually show much interest. As ovulation approaches and the female's sex skin reaches maximum size, male competition heats up, and the alpha male usually wins. He and the female form a "sexual consortship" in which the male closely follows the female and they groom and mate at high rates. Depending on the male's rank, other males may challenge the consorting male and attempt to take over his female. It is an anxious, stressful time for the consorting male, and his glucocorticoid levels may rise significantly (Bergman et al. 2005). To reduce the risk of a challenge from a rival, the consorting pair may move to the edge of the group, away from others. Soon after ovulation, the female's sex skin deflates rapidly and the consortship usually ends, though the female continues to mate with other males for several more days.

This is not to say that the alpha male has a total monopoly on all matings. Often, more than one female is cycling simultaneously, so that second- and third-ranking males are also able to form consortships. There is also often a considerable amount of "sneaky" mating. Paternity data derived from fecal samples indicate that, although the alpha male fathers the most offspring, other males also achieve some reproductive success (Altmann et al. 1996).

The particular combination of males in the group at any given time sets the overall tone. When there are several young, ambitious strivers jockeying for the alpha position, wahoo contests and fights occur several times a day. Ranks change often, and glucocorticoid levels are high, particularly among males at the top of the hierarchy, who have the most to lose. Instability in the dominance hierarchy is the single most important cause of stress in male baboons (Sapolsky 1992, 1993, 2005; Bergman et al. 2005). If, on the other hand, there is one clearly dominant alpha male and few rising natal or immigrant males, life is much quieter. Displays are rare, chases even rarer, ranks remain stable, and animals spend their day peacefully foraging, grooming, or sitting in the shade. High-ranking males, whose control is not at risk, exhibit low glucocorticoid levels. Now the males who experience the greatest stress are those at the bottom of the hierarchy (Sapolsky 1993; Alberts et al. 2003; Bergman et al. 2005).

Immigrants, infanticide, stress, and male-female friendships

Whatever the mood in the group at the time, the arrival of a new immigrant male disrupts everything. Youngsters run to inspect the newcomer, boldly racing up to him, staring, and then quickly whirling around to present their rumps as a sign of submission. Resident adult males are more seriously upset. They become vigilant and restless, warily tracking the new male's every move but avoiding any direct confrontation. They seem to be assessing him from a distance, waiting for a wahoo or yawn that will reveal his fighting ability.

The immigrant, of course, is making similar assessments. He appears to recognize immediately which are the high-ranking males, and although he avoids these individuals, he may confront lower-ranking males directly, by walking deliberately at them. He may also solicit friendly interactions from females and juveniles—particularly the group's alpha female—by grunting and lipsmacking at them as he approaches. (Lipsmacks are a universal signal of friendly intent in monkeys. In baboons, as we discuss in later chapters, grunts also signal benign intent.) Cycling females may respond to the new male's solicitations by presenting to him.

The most dramatic reactions to a new immigrant, however, come from lactating females, who raise their tails, grab their infants, and race away screaming whenever he begins to approach. Even his grunts fail to reassure them. Their anxiety is well founded, because most immigrant males who rise to the alpha position commit infanticide (Palombit et al. 2000; Cheney et al. 2004). Infanticide has also been observed or strongly suspected when a new immigrant male has arrived in the group and risen in rank, but not attained the alpha position. Infanticide, as we mentioned in Chapter 3, is the single most important cause of infant mortality. At least 53% of all infants born during our study have died as a result of confirmed or suspected infanticide (Fig. 12).

Why do males commit infanticide? The now well-documented answer originates from research conducted by Sarah Hrdy (1977) on Indian langur monkeys. It has since been supported by many other studies of monkeys, rodents, carnivores, and other mammals (Janson and van Schaik 2000). Taking baboons as an example, consider the following

Figure 12. The immigrant alpha male Dougie bites the infant he has just killed. Photograph by Ryne Palombit.

facts. Although the alpha male monopolizes most of the matings with fertile females, he can expect to hold the alpha position for only seven or eight months before a new immigrant defeats him. Furthermore, because the gestation period in baboons is six months and lactating females nurse their offspring for at least a year before resuming sexual cycling, the majority of females will be unavailable for mating during his brief tenure. The male, therefore, has two choices: he can accept the situation as it is and mate with the few females who are currently cycling, or he can kill as many infants in the group as possible, thereby hastening the resumption of lactating females' sexual activity. In so doing, he will have replaced another male's infant with his own.

Infanticide, therefore, is a sexually selected trait that enhances a male's fitness, though at the expense of the fitness of one or more females. Not surprisingly, it occurs most often in sexually dimorphic, polygynous species, where male competition is intense and females find it difficult to defend themselves against a male's attacks. This is not to say that males understand why infanticide is adaptive, or that they explicitly calculate their expected reproductive success when they commit infanticide. Similarly, not all immigrant males who achieve alpha status commit infanticide. Infanticide appears to be an opportunistic strategy whose occurrence depends both on the number and personalities of resident males and on the personality and aggressiveness of the immigrant. Indeed, the proximate mechanisms that underlie a male's decision to kill an infant remain elusive.

Data on stress confirm that, for female baboons with infants, the arrival of a high-ranking immigrant male is a period of great anxiety. Research by Jacinta Beehner, Thore Bergman, and Anne Engh has shown that rank changes among resident males produce no signs of stress in females, even when a long-time resident or natal male rises to the alpha position. This is probably because resident and natal males pose little infanticidal risk. When a new immigrant arrives, however, female glucocorticoid levels rise significantly, but only among those who are lactating. Cycling and pregnant females show little or no response. And if the new immigrant rises to the alpha position and begins to commit infanticide, lactating females' glucocorticoid levels rise even higher (Beehner et al. 2005; Engh et al. 2006a).

Even though they are much smaller than males, female baboons defend their infants vigorously against real and potential threats, sometimes leaping on and biting any male that attempts to attack them. Many males, though, are relentless in their pursuit, stalking females for months and waiting for an opportune moment to grab the infant in a drive-by attack. Most lactating females, therefore, adopt an alternative strategy: they form a close "friendship" with a resident adult male. Although lactating females form friendships with males of all ranks, the most attractive friends, and those over whom females compete the most, are males who were in the alpha position when the females' infants were conceived (Palombit et al. 2001). These males not only provide the best protection, but they are also most likely to be the infant's father. They are therefore highly motivated to protect it. In fact, male friends go to considerable lengths to protect infants, often grabbing and carrying infants when approached by a potentially infanticidal male.

Friendships are the strongest and most enduring bonds formed between males and females (Fig. 13). While sexual consortships last for only several days, friendships persist for as long as a year. Friends forage and sit together, and groom each other at high rates. Males also baby-sit when their friends are foraging, tolerating outrages against their dignity like tail-pulling, back-sliding, and head-sitting.

Our colleague Ryne Palombit has amassed a variety of evidence supporting the view that friendships in baboons are a counterstrategy against infanticide (Palombit et al. 1997, 2000). To begin with, friendships are most common when there is a new immigrant male present in the group, and especially common when the new immigrant achieves

Figure 13. Females often establish a close friendship with a male after the birth of an infant. Photograph by Roman Wittig and Cathy Crockford.

alpha status. At other times, lactating females may form no friendships at all. When friendships are formed, they are largely the work of the female, who follows her friend and remains as close as possible to him. The female's greater initiative is not surprising, because she and her infant clearly stand to benefit from the friendship. Playback experiments have shown that males recognize the screams of their female friends and respond preferentially to their cries of distress. Perhaps as a result, when a potentially infanticidal male arrives in the group lactating females with male friends experience significantly lower glucocorticoid levels than those without friends (Beehner et al. 2005; Engh et al. 2006a). Having a protector clearly reduces stress.

Because each infant represents a large proportion of a female's lifetime reproductive success, females may derive a greater benefit from a friendship than males. As long as there is some probability that the infant he is defending is his own, however, friendships also benefit males. Friendships, therefore, are opportunistic and self-serving. They exist primarily because they offer protection to infants who may be the male's offspring. Once an infant dies or is weaned, the friendship ends abruptly (Palombit et al. 1997). The male may continue to interact with the female, but the close proximity and grooming that were once so evident largely disappear. Males do, however, continue to defend the now independent juvenile, and if the juvenile is orphaned the male will often take over the role as the juvenile's primary adult companion. Indeed, over time a male's successive friendships transform him into a Pied Piper, and he is accompanied wherever he goes by a troupe of tumbling and twirling youngsters. For an old, slow-moving, low-ranking male, it is as good a way as any to finish out his years.

FIVE

Females: Kinship, Rank, Competition, and Cooperation

Three or four families in a country village is the very thing to work on, and I hope you will ... make full use of them while they are so very favorably arranged.
JANE AUSTEN, 1814: LETTER TO HER NIECE ANNA

It is a truth universally acknowledged in baboon society that, while success in the male world is determined by sex, fighting, and posturing, success in the female world depends on family, social networks, and intrigue.

Kinship

The social structure of female baboons is very similar to that of many other Old World monkey species, including vervets and macaques. Females live throughout their lives in their natal group and maintain close, enduring bonds with their matrilineal kin (Silk et al. 2006a,b; see Appendix for the study group's family trees). These relationships are best observed through grooming. Adult females are assiduous groomers and can spend four or five hours a day grooming with several different partners. Over the course of a year, a female may have dozens of grooming partners, including offspring, other relatives, adult males, and unrelated adult females. The great majority of her grooming, though, is exchanged with a few of her close female rela-

Figure 14. Luxe forms a grooming chain with her daughters Bex and Naxos. Photograph by Anne Engh.

tives (Fig. 14). Grooming involves far more than just the removal of ec-toparasites. When two animals groom, their behavior not only provides immediate satisfaction but also strengthens the bond between them. This not only causes them to groom each other again but also reinforces their tendency to spend time together, tolerate each other at feeding sites, huddle together on cold days, and support each other in aggres-sive alliances (Seyfarth and Cheney 1984, 1990; Silk et al. 1999, 2004, 2006a,b; Schino et al. 2003). All of these interactions forge a relation-ship that is more than the sum of its parts.

Close bonds among matrilineal kin have evolved through kin selec-tion, the evolutionary process that favors individuals who can recognize their close genetic relatives and assist in their reproduction (Hamilton 1964). Kin selection creates nepotistic biases in the behavior of animals as diverse as ants and elephants. In baboon society, the females most likely to cooperate are close matrilineal kin, who have both a genetic predisposition to do so and close social relationships that serve as a mechanism for identifying kin. (Recent research on rhesus macaques and baboons in Kenya [Widdig et al. 2001; Smith, Alberts, and Altmann 2003] has revealed that cooperation can also occur among patrilineal kin, who also have a genetic reason to do so but for whom the mechanisms underlying recognition are as yet unknown. For ease of discussion, in this book we use the terms "kin" and "relatives" to refer to maternal relatives.)

Alliances play a special role in baboon metaphysics. An alliance occurs when two animals become involved in a fight and one of them recruits a bystander to join her as an ally. Here is a typical example, from data collected in 1992. Sierra, a juvenile female at the time, was walking toward a jackalberry tree where other animals were feeding. In her path sat Jeanette, a low-ranking adult. Contrary to her usual behavior, Jeanette did not move off at Sierra's approach; instead, she threatened Sierra with a quick bob of her head. Sierra immediately lunged at Jeanette, giving a series of threat-grunts, and, for good measure, a loud scream. Like many other vocalizations in the baboons' vocal repertoire, threat-grunts and screams are individually recognizable (Owren et al. 1997; Palombit et al. 1997; Rendall et al. 1999; Bergman et al. 2003). As a result, Sierra's calls not only challenged Jeanette with the threat of reprisal but also broadcast a signal to her own matrilineal relatives. Immediately, Sierra's mother, Sylvia, and her aunts Selo and Stroppy came running to Sierra's aid. Together the four leapt on Jeanette, pinned her to the ground, and bit her tail. Jeanette dropped her food and ran away screaming, but no one responded to her calls.

To succeed in baboon society, a female must, of course, recognize her own relatives. But to participate in coalitions, she needs more than this. Because baboons and other monkeys form most of their alliances with close kin, it also behooves a female to take a non-egocentric perspective and recognize *other individuals'* social networks. Alliances involve not just the support of one individual but also hostility toward another— and often, by extension, the other animal's family. They require nuanced decisions about when to join, whom to join, whom to threaten, and whom to ignore. In other words, life in a baboon group seems to require an understanding of the group's entire kinship system. We consider this question further in Chapter 6.

Rank

Like males, female baboons form linear, transitive dominance hierarchies. There, however, the similarity ends. Whereas male dominance ranks are acquired though aggressive challenges and change often, female ranks are inherited from their mothers and remain stable for years at time. Furthermore, most female dominance interactions are very subtle. Although threats and fights do occur, they are far less common and violent than fights among males. Instead, most female dominance interactions take the form of supplants: one female simply approaches another and the latter cedes her sitting position, grooming partner, or food. The direction of supplants and aggression—and the resulting female dominance hierarchy—is highly predictable and invariant. The alpha female supplants all others, the second-ranking supplants all but the alpha, and so on down the line to the 24th- or 25th-ranking female, who supplants no one.

The female dominance hierarchy in most Old World monkeys is in fact a hierarchy of matrilines. Daughters acquire ranks similar to their mothers', and sisters assume adjacent ranks. In marked contrast to males, however, high-ranking females are not necessarily in better condition than low-ranking ones. Even though the highest-ranking female might be old and decrepit, and the second-ranking female a spindly adolescent, all others defer to them.

Female baboons, like female macaques and vervet monkeys, acquire and maintain their dominance ranks by processes that are both physical and psychological. Infant baboons are extremely attractive to adult and juvenile females, who gather around to touch, hug, and examine them. But while all infants are attractive, the babies of low- and high-ranking mothers receive somewhat different treatment. If a female wants to handle an infant whose mother ranks lower than she does, she simply walks toward the mother, grunts repeatedly to appease her (low-ranking mothers are often very nervous and grunts seem to reduce their anxiety; see below) and touches or hugs the infant (Fig. 15). The approaching female is not at all aggressive, but there is no hint of deference in her behavior. By contrast, if a female wants to handle an infant whose mother ranks higher than she does, the process is considerably more elaborate and obsequious. The female approaches the mother tentatively, grunting several times while gazing at the infant. If the mother does not threaten her away, the female moves closer, then sits and grunts again. If there is still no threat from the mother, the female inches her way to within arm's length, and while furiously grunting to

Figure 15. Bex, the juvenile daughter of the high-ranking Luxe, examines the infant of the much lower-ranking CP. Photograph by Anne Engh.

the mother reaches out and gently touches the baby. Even before they begin to move independently, high-ranking infants are treated with a caution and deference that low-ranking infants never receive.

We say that rank is acquired and maintained by both physical and psychological processes because high-ranking baboons are protected by both the actual, physical intervention of their kin and their kin's sycophants and the implied threat of such intervention. In baboons and other monkeys whose social organization is built around nepotistic rank acquisition, high-ranking females are not more aggressive than lower-ranking females, nor are high-ranking families always larger than low-ranking families. Nevertheless, high-ranking individuals consistently receive the most support from both their kin and unrelated animals (reviewed by Walters and Seyfarth 1987; Chapais 2001; Silk 2002). Low-ranking females rarely unite to form an alliance against a higher-ranking female. Instead, females form alliances with individuals who already rank higher than their opponent. Most alliances, therefore, are gratuitous and redundant: they do not affect the outcome of the dispute, but function only to preserve the status quo (see Chapais 2001 for experiments testing this observation in Japanese macaques). The result is a conservative society in which females acquire their family's rank, and high-ranking juveniles and adults maintain their status because they are—well, high-ranking.

Although alliances in baboons and many other monkeys play a crucial role in the acquisition and maintenance of female rank, females form alliances against other females at surprisingly low rates. Whereas females intervene in many of their juvenile offspring's disputes, and female vervets and macaques also form alliances against males at relatively high rates (the great disparity in size between male and female baboons probably precludes female baboons from forming alliances against males), fewer than 5% of females' disputes with other females result in alliances (Wittig et al. 2007b). The frequency in the Okavango is 4%, comparable to the frequency found in other areas of Africa (Silk et al. 1999, 2004).

This low rate of female-female alliances seems puzzling at first glance, because the linear hierarchies in these species are thought to depend on familial alliances. Female baboons and other monkeys, though, are very vocal animals, and otherwise uninvolved bystanders often give threat-grunts while observing other females' disputes, as if indignantly reproaching the tiff from the sidelines. These simian Madame Defarges are usually close relatives of the higher-ranking combatant, and they are almost always higher-ranking than at least one of the combatants. Because nothing stops the spectators from joining the scrum, their threat-grunts may function as implied threats (Fig. 16). In the Okavango baboons, roughly 6% of females' disputes result in "vocal alliances" by female bystanders. Taken together, physical alliances and the vocal threat of an alliance may be sufficient to maintain a stable and conservative matrilineal hierarchy.

Vocal alliances are low-cost signals that announce the signaler's willingness to intervene physically if the dispute is not settled quickly. As we discussed in Chapter 4, theory predicts that animals should always attempt to settle disputes through low-cost displays that allow contestants to assess each other's competitive ability and likelihood of support before a fight escalates and results in injury (Maynard Smith 1982;

Figure 16. A juvenile female enlists the support of an adult female while threatening a younger juvenile from a lower-ranking family. Photograph by Adrian Bailey/Aurora.

Bradbury and Vehrencamp 1998). We might therefore expect vocal alliances to be widespread in animals. In the next chapter, we describe a playback experiment designed to test whether baboons interpret bystanders' threat-grunts as alliance support.

Like ranks among families, dominance ranks within baboon families are clear and unambiguous. They follow the rule of youngest ascendant: sisters assume ranks in inverse relation to their ages, with the youngest daughter ranking higher than all of her sisters. The process of rank acquisition within monkey families is not well understood, but younger daughters probably acquire ranks above their sisters at a very early age, when the mother consistently intervenes on the side of her smallest, most vulnerable offspring (reviewed by Chapais 2001; Silk 2002). In some families, one or more daughters may eventually rise in rank above her mother, especially as the mother ages and her reproductive value decreases (Combes and Altmann 2001). In others, the matriarch never cedes her position.

The relative ranks of females are extraordinarily stable. When we began our study in 1992, the seven matrilines in the group had already held their relative positions for many years. In 1992, the alpha female was Stroppy, the second-ranking female was her daughter Sylvia, and the third-ranking female was Beth. Stroppy had two juvenile daughters,

Selo and Swallow, and Beth had one juvenile daughter, Luxe. In 1994 Stroppy died, and her daughter Selo assumed the alpha position. As of June 2006, Selo still occupied the alpha position. Palm, her daughter, was second-ranking; Swallow, Selo's sister, third-ranking; and Sylvia, Selo's older sister, fourth-ranking. Beth's daughter Luxe ranked fifth, and Luxe's daughter Bex ranked sixth. Although many of the individuals have changed, the same two matrilines have occupied the top two positions for decades.

Challenges to rank

Baboon society is highly nepotistic, but it is not ruthlessly despotic like the societies of wolves (*Canis lupus*) or dwarf mongoose (*Helogale parvula*), for example, where only the dominant female breeds. In baboon groups, even low-ranking females are able to raise offspring successfully. Baboon society is also highly conservative, as we have mentioned: females typically support the higher-ranking of the two opponents. Perhaps because the benefits to be derived from high rank do not usually outweigh the potential costs of a serious fight, overt challenges to the existing hierarchy are rare. Although a low-ranking female will unhesitatingly cuff a high-ranking juvenile who is attempting to take her food, she will do so only after ensuring that none of the juvenile's relatives are within earshot. Nonetheless, beneath the peaceful, orderly hierarchy lurk individuals—and indeed whole matrilines—just waiting for an opportunity to disrupt the social order.

One such opportunity seems to have arisen in July 2003, when Leko, the matriarch of the fourth-ranking matriline, innocuously began a sexual consortship with Loki, a middle-ranking male. For reasons still unclear, adolescent females from the fifth-, second-, and third-ranking matrilines responded with indignation. They threatened Leko with head bobs, threat-grunts, chases, and bites. Leko and her daughters, Lizzie and Lissa, responded in kind, but Leko was soon driven to the periphery of the group. The bulk of the attacks involved members of the fifth-ranking matriline—Balo, her daughters Amazon and Domino, and her sister, Atchar. For a week the members of both matrilines, as well as females from other matrilines, fought often and sometimes violently. After a few days Leko's and Loki's consortship ended, and the Leko family's retaliation began in earnest. Lissa seemed particularly incensed by the pretenders' challenge. One morning she was able to isolate Balo, the two tussled in a violent scrum, and Balo received a bloody wound on her eye. Gradually, the attacks abated, and everyone reverted to her former rank.

These attacks were completely unexpected, for two reasons. First, females typically ignore one another's sexual consortships, both because they probably fear the male's intervention on behalf of his partner and because most adult females are usually pregnant or busy with young offspring. The attack on Leko might have occurred because at the time the group contained an unusually large numbers of adolescent and cycling females with no offspring who may have regarded Leko's consortship as mate competition. Second, the attacks were surprising because they violated two of the rules that govern female dominance relations: Leko was threatened by females who ranked lower than she, and these low-ranking individuals managed to recruit alliance partners who ranked higher than Leko.

Later that year, a second challenge to the existing hierarchy was more successful. At the time, the group's third-ranking matriline was unusually large, and included the eight surviving daughters, granddaughters, and great-granddaughters of the venerable Sadie, an aged female who had died in 1994 and was justifiably referred to as "the mother of all baboons." Perhaps because this matriline was so large, it had become less cohesive; most of its members now interacted primarily with their immediate family members, ignoring more distant kin. Whatever the reason, a coup was again instigated and Cat, one of Sadie's adult granddaughters, was the first to fall. Again, the ostensible cause was a sexual consortship.

After separating herself for several days during her consortship with the alpha male, Cat was attacked and driven to the periphery of the group when she attempted to return. After enduring repeated attacks for a week, Cat was eventually allowed to rejoin the group, but other females continued to threaten and chase her. None of Cat's sisters, aunts, or nieces interceded on her behalf, and one of her sisters participated in the threats against her. CP, the lowest-ranking female with no close relatives, took advantage of Cat's lack of social support and harassed Cat mercilessly. Initially, CP threatened Cat only when she was able to recruit high-ranking females against her. Within a week, though, CP and all the other lower-ranking females were able to dominate Cat individually. Cat and her one-year-old daughter fell to the bottom of the hierarchy. During the next three months, Cat's aunt, Helen, her sister, Hannah, and Hannah's two daughters were also forced to the bottom of the hierarchy by females from lower-ranking matrilines. None of the females defended themselves when threatened, and none of their relatives intervened on their behalf. Another of Cat's sisters and two nieces maintained their ranks. Prior to this event, the female dominance hierarchy had remained stable for over 20 years.

Not surprisingly, these rare rank upheavals were stressful for the individuals involved. During both periods of instability, females whose ranks were potentially affected showed significant increases in glucocorticoid levels, while females in higher-ranking, unaffected matrilines did not (Engh et al. 2006a). Interestingly, glucocorticoid levels in matrilines that fell in rank were no different from those in matrilines that rose. Apparently, rank instability per se was more stressful than its outcome.

Social striving and the interaction between kinship and rank

William Thackeray's *Vanity Fair* describes the ascent of Rebecca Sharp, a ruthless orphan who abandons all scruples in her drive to achieve the highest social status possible.

And now ... it became naturally Rebecca's duty to make herself, as she said, agreeable to her benefactors, and to gain their confidence to the utmost of her power. Who can but admire this quality of gratitude in an unprotected orphan; and, if there entered some degree of selfishness into her calculations, who can say but that her prudence was perfectly justifiable?

Low-ranking female baboons are all Becky Sharps. Ruby, for example, was born in 2000 and orphaned in 2003, when her mother and infant sister were killed in an apparent leopard attack. Ruby's mother had been a member of the fourth-ranking matriline, ranked 17th in a hierarchy of 23 females. Although Ruby was left without any close female relatives, she had an older brother, Rick, and with his support was able to maintain her mother's rank.

Eight months later, though, Rick emigrated from the group. At this point, Ruby apparently decided to seek her fortunes elsewhere. Unimpeded by the anchor of a low-ranking mother, Ruby befriended the members of Selo's family, the highest-ranking matriline. Her particular focus of attention was Sylvia, Selo's ancient sister and still, at 22, one of the meanest females in the group. Ruby groomed Sylvia regularly; occasionally Sylvia reciprocated. Ruby also spent much of her time feeding near Sylvia, so that when other baboons tried to threaten her away from food Sylvia was often nearby. Perhaps because of Ruby's grooming, or perhaps because Sylvia mistakenly believed that some of these threats were directed at her (unthinkable!), Sylvia sometimes chased Ruby's aggressor. By the time Ruby reached adolescence, she had ascended above the third-ranking matriline. Her relationship with Sylvia was not as stable or enduring as a close bond between kin, but—however temporary it may have been—it served a purpose.

Ruby's ascent emphasizes that matrilineal kin are sometimes more of a liability than an asset, at least when it comes to rank. A female's rank depends on her family's status, and for a low-ranking female this can be an impediment. If an orphaned juvenile female has an older brother but no sisters, she invariably assumes a rank higher than she should. Given the enormous advantage that males enjoy in size and aggressiveness, a brother's support can be useful indeed. But if the female has even one older sister, she almost always retains her mother's rank. Apparently, the presence of even one female relative serves as a sufficient reminder to other females of her appropriate station. On the other hand, a female relative is not a complete liability, because juvenile females who lose all of their relatives sometimes assume ranks far below their mother's. Thus, the acquisition and maintenance of rank for members of low-ranking matrilines is often more fluid, subtle, and unpredictable than it is in high-ranking matrilines, if only because the status quo is less satisfactory.

Although a female might sometimes gain from distancing herself from her matrilineal kin, many female relatives are always better than none. Females form their strongest bonds with their matrilineal kin, and females who lose their relatives must strive to form bonds with others. Research on baboons in Amboseli National Park, Kenya, has shown that socially integrated females enjoy higher infant survival than females with weaker social networks (Silk et al. 2003; Chapter 6 of this volume). Furthermore, as we mentioned in Chapter 3 and discuss further below, females who lose a close relative to predation experience a significant increase in stress. This stress can be alleviated through bonds with other individuals (Engh et al. 2006b). Clearly, the presence of female kin buffers female baboons in many ways. At the same time, however, high-ranking females make good allies. It therefore behooves individuals like Ruby to curry favor with the members of high-ranking matrilines. Numerous studies have now shown that the distribution of grooming among female monkeys is influenced by both the attraction to kin and the attraction to high-ranking individuals (Seyfarth 1977; Fairbanks 1980; Schino 2001; Ventura et al. 2006).

How do low-ranking female monkeys benefit from grooming high-ranking females? Because such interactions seldom result in rank changes or a close, enduring bond, the immediate benefits are not always clear (Fig. 17). Female baboons, macaques, and New World capuchin monkeys (*Cebus capucinus*) are most likely to groom unrelated females who groom them (Silk et al. 1999, 2006a,b; Ventura et al. 2006; Manson et al. 2004), but the benefits to low-ranking females are not limited to grooming received in return. In some groups, low-ranking females are more likely to groom higher-ranking females who threaten them at high rates, suggesting that high-ranking females use grooming as a form of extortion (Silk 1982; Perry 1996; Barrett et al. 2002; Schino et al. 2005). In other groups, the opposite pattern is evident. Female vervet monkeys form more alliances with unrelated females with whom they groom often (Seyfarth 1980; Cheney and Seyfarth 1990). Playback experiments have also shown that females respond more strongly to the recruitment calls of an unrelated female if they have recently groomed with that female than if no grooming has taken place

Figure 17. Although females groom their relatives at the highest rate, they also groom nonrelatives. Jackalberry grooms Amazon, a female from a higher-ranking matriline. Photograph by Anne Engh.

(Seyfarth and Cheney 1984). Similarly, high-ranking female long-tailed macaques (*Macaca fascicularis*) are significantly more likely to support a lower-ranking female in a fight if the lower-ranking female has recently groomed them than if she has not (Hemelrijk 1994). Across many primate species there is a significant positive correlation between grooming and support in alliances (Schino 2007), but the causal mechanisms that underlie this correlation are complex. Depending on the particular social group and the individuals that comprise it at the time, grooming may function to forge a closer relationship, enhance alliance support, or to appease aggressive dominants (Schino et al. 2005).

Regardless of the long-term benefits that a low-ranking female baboon may derive from a grooming interaction with a higher-ranking female, it is clear that she regards it as a signal of friendly intent. In playback experiments, low-ranking females respond weakly if at all to the threat-grunts of a high-ranking female with whom they have recently groomed, suggesting that they do not expect to be threatened by her. They seem to equate grooming with a low probability of aggression. We discuss these playback experiments in detail in Chapters 6 and 8.

In sum, monkey society is governed by the same two general rules that governed the behavior of women in so many 19th-century novels: stay loyal to your relatives (though perhaps at a distance, if they are a social impediment), but also try to ingratiate yourself with the members of high-ranking families. The two rules interact in interesting ways. For the members of high-ranking matrilines, the rules of kin-based and rank-based attraction reinforce one another, whereas for the members of low-ranking families they counteract. A member of a high-ranking matriline is attracted to her kin not only because they are members of the same family but also because they are high-ranking. A member of a low-ranking family may be attracted to her kin, but she is also drawn away from them by her attraction to unrelated, higher-status individuals. As a result, high-ranking families are often more cohesive than low-ranking ones (e.g., Fairbanks 1980; Silk et al. 1999). Or, to paraphrase Tolstoy's *Anna Karenina,* all high-ranking families are alike in their cohesiveness; each low-ranking family is cohesive or not, in its own way.

A few years ago, a member of the British royal family visited us in the field and spent a morning following the baboons. On being told the details of the baboons' inherited, rank-based society she became both excited and relieved, as if a longstanding dilemma had at last been resolved and an onerous weight lifted from her shoulders. "I always knew," she declared, "that when people who aren't like us claim that hereditary rank is not part of human nature, they must be wrong. Now you've

given me evolutionary proof!" Shortly thereafter she returned to her en-
tourage, spirits uplifted, leaving us to ponder the wider implications of
our work.

Cooperation and reconciliation among unrelated females

For readers who are not members of hereditary royal families, the rank-
based society of baboons begs several questions: Why do females care
whether or not they are high-ranking? Furthermore, if there is a big
advantage to being high-ranking, why do low-ranking females remain
in the group at all? Why not start their own group, with themselves and
their family at the top? There are several answers.

First, for baboons and other monkeys, group life is essential. Living in a group enables females both to defend food resources against
other groups and to reduce the probability of predation (Wrangham
1980; van Schaik 1983; Sterck et al. 1997). Second, life for a low-ranking
female baboon is not all that bad. In some mammalian species—like
wolves, wild dogs, and some mongooses—there is strong reproductive
skew among females, and only the most dominant individual breeds.
This is not true in monkey societies, where females of all ranks produce offspring. True, high-ranking females enjoy priority of access to
food resources, often begin their reproductive careers at younger ages,
and have slightly shorter intervals between births (for reviews see Silk
2002; Cheney et al. 2004). Over several generations, these small benefits can translate into a significant reproductive advantage (Altmann
and Alberts 2003). For baboons in the Okavango, however, such small
rank-based advantages are overridden by infanticide and predation, two
factors that strike high- and low-ranking females with equal force and
are far more important in determining their lifetime reproductive success (Cheney et al. 2004).

Indeed, perhaps because all baboons benefit from group life, high-
ranking females often seem to go out of their way to groom low-ranking
females, handle their infants, and hug or touch them while foraging.
Grunts play a particularly important role in these interactions. Grunts
are the most common vocalization given by baboons, and they occur at
a very high rate. They are tonal, individually distinctive, and function
in many different ways to facilitate social interactions (Chapter 10).

When a high-ranking female approaches a lower-ranking one, there
is always some ambiguity about what will happen next (Silk et al. 2000).
On the one hand, the high-ranking female might be attempting to

Figure 18. A juvenile female from a high-ranking matriline attacks a low-ranking mother who is attempting to prevent her from handling her infant. Physical attacks like these are relatively uncommon among females. Photograph by Adrian Bailey/Aurora.

groom the lower-ranking one or handle her infant. On the other hand, she might be attempting to supplant or threaten the lower-ranking female from the sausage fruit or palm nut that she is eating (Fig. 18). This uncertainty creates a dilemma, both for high-ranking females who want to behave in a friendly manner and for low-ranking females who are reluctant to give up their food or shady spots. In stable social groups where individuals interact frequently with one another, such uncertainty should favor the evolution of reliable, honest signals that provide accurate information about the signaler's probable behavior (Silk et al. 1996; Maynard Smith and Harper 1995). In baboons, vervets, and macaques, grunts serve this function: they are highly predictive of friendly behavior, and listeners respond accordingly. If a high-ranking female baboon grunts while approaching a lower-ranking one, the lower-ranking female is likely to remain seated. Conversely, if the approacher remains silent, the lower-ranking female usually moves away (Cheney et al. 1995b; Silk et al. 1996, 2000). As in the case of grooming, high-ranking females almost never threaten a lower-ranking female after grunting to her (though there are exceptions, as we discuss later).

Grunts also function to reconcile opponents after a dispute. Nonhuman primates are often aggressive toward one another, yet they live in relatively stable, cohesive social groups. A number of studies have shown that opponents mollify the effects of aggressive competition by grooming or interacting in a friendly manner soon after they fight (Kappeler and van Schaik 1992). In baboons, such interactions are usually preceded by a grunt from the dominant aggressor to her subordinate opponent (Cheney et al. 1995b; Silk et al. 1996).

Testing the reconciliatory function of grunts

When two female baboons get into a squabble, the higher-ranking animal may lunge at, hit, chase, or even bite her opponent. Typically the subordinate cowers, screams, and bares her teeth in a fear grimace. In roughly 13% of all fights, however, the dominant female grunts to her victim shortly after the fight ends. The effect on her opponent is striking. She stops cowering and often stops moving away. She seems immediately to relax. It is as if the grunt has somehow conveyed the message "Don't worry; I'm not going to threaten you again."

But reconciliation is a loaded term. In its richest interpretation it implies that the dominant female recognizes that her victim is anxious and afraid—emotions that the dominant female herself is certainly not experiencing—and grunts to alleviate the subordinate's anxiety. The subordinate, in turn, recognizes the dominant's intent to reconcile. We will leave discussion of this loaded, mentalist interpretation to Chapter 8, and focus here on a simpler problem: how can we be sure that it is the grunt, and not some other factor, that causes the subordinate to change her behavior? Perhaps it is something about the dominant female's posture, the nature of the fight, or the two females' past history of interactions that causes what appears to be reconciliation. Furthermore, even if we can determine that the grunt is, indeed, the crucial reconciliatory signal, how can we be sure that it serves this function by signaling to the subordinate something about her opponent's behavior? Perhaps hearing the grunt of *any* dominant female is sufficient to cause the subordinate to relax.

To answer these questions we need somehow to probe the subordinate's mind in the minutes after she has fought with the dominant. Like intrusive journalists, we need to enter the scene just after the fight and ask the subordinate "What was it that made you relax? What would you have done if your opponent *hadn't* grunted?" Playback experiments provide one means to do this. Because these are the first of many playback experiments we will discuss, we describe them in some detail here.[1]

* * *

Playback experiments can be designed to simulate normal social interactions or to present subjects with an anomalous sequence of calls that mimics an event they have never experienced. Whatever their exact purpose, most follow the same general protocol. We first create a recorded archive of calls given by known individuals. Next, having calibrated a recording to match the amplitude and duration of naturally occurring vocalizations of that type, we play the call to a predetermined subject from a battery-powered loudspeaker concealed in tall grass or behind bushes or trees. As much as possible, we try to ensure that the loudspeaker is placed in the same general direction as the individual whose call we are about to play and that both the signaler and any other relevant animals (like her close relatives) are out of sight and earshot.

As soon as the speaker is properly situated and the subject's head is oriented in a direction at least 90° from the speaker, we conduct the playback trial, filming the subject for a predetermined time before and after the call is played. This record allows us to measure the subject's latency and duration of looking in the direction of the speaker, as well as any other behavioral responses. We then follow the subject for a predetermined time after playback, noting whom she approaches and with whom she interacts. Most of our experiments follow a within-subject design. In some paired trials, the same individual hears two different calls mimicking two different social interactions played to her in the same context. In others, the subject hears the same call played to her in two different contexts—for example, after two different social events.

---

1. The first playback experiment with monkeys appears to have been conducted by Richard L. Garner in the Central Park Zoo in 1891. Using a phonograph, he played back the "salutations" of the local group of rhesus macaques to a shipment of new monkeys, who responded with great excitement. The technique was subsequently adopted by a variety of investigators, some of whom found that it enhanced their hallucinatory experiences. Marianne Faithfull, the 1960s pop star and companion to the Rolling Stones, reports that in 1967 Brian Jones played the soundtrack of a song he had composed to a group of Barbary macaques living on the Rock of Gibraltar. "We approached the troop of monkeys very ceremoniously. Bowed to them and told them we were going to play them some wonderful sounds. They listened to all this very attentively, but when Brian turned on the tape recorder, they didn't seem to care for it. They seemed alarmed by it and scampered away shrieking. Brian got very upset. He took it personally. He became hysterical and began sobbing" (Faithfull and Dalton 1994).

In most experiments, our aim is to compare test sequences against at least one control. As much as possible, we randomly vary the order in which subjects hear control and test sequences.

Previous studies have shown that subjects' responses to call playbacks differ considerably according to the type of call they hear. Some call types—most notably alarm calls—evoke qualitatively different responses, like running into trees or looking up (e.g., Seyfarth et al. 1980). Others elicit only a look toward the speaker. In these cases, differences in responses to different playback stimuli are measured in terms of the duration that subjects look toward the speaker, a quantitative rather than a qualitative measure. There is a long and important tradition of psychological experiments that use quantitative differences in orienting to detect differences in responses to signals. In studies of nonhuman primates, this dependent variable has revealed consistent, significant differences in subjects' responses to acoustically different calls, to calls from different individuals, and to the same call played back in different contexts (reviewed in Seyfarth and Cheney 1997a).

To retain their credibility, playback experiments must be conducted at low rates and blend in as naturally as possible with the baboons' daily life. With the exception of trials involving anomalous sequences, we play calls at less than a tenth of the rate of their natural occurrence, and trials involving the same subject are always separated by at least two days. Because opportunities to conduct a playback experiment are often rare and fleeting, the actual rate of experimentation is far lower than this. It is not at all unusual for one series of trials to take a year to complete. Finally, many of our attempts to conduct a playback trial are thwarted when some piece of equipment perversely ceases to function, an elephant or other unwelcome intruder suddenly appears, or our subject wanders off to interact with precisely the individuals we want her to avoid. As a result, the baboons have many opportunities to see us carrying loudspeakers and video cameras without hearing a call.

Despite all of these precautions, none of our experiments ever achieves the precision and control of many laboratory tests. We simply do not know everything that has happened to our subjects on the day they are tested, nor can we control the myriad contextual variables present under natural conditions. Many of these problems can be alleviated by allowing different trials to serve as each others' controls. If some aspect of our protocol is inadvertently biased, then it should be equally biased across different trials. In the end, we control what we can and hope that the benefits of experimenting on animals in their natural habitat outweigh the imprecision of our methods.

...

In a first attempt to test the reconciliatory function of grunts, we and Joan Silk designed an experiment in which we played the tape-recorded scream of a dominant female (say, seventh-ranking Hannah) to a lower-ranking female (say, 12th-ranking Nimi) under three different conditions: first, after the two females had fought and Hannah had grunted to Nimi; second, after the two females had fought and separated in silence; and third, after no fight had recently occurred. In playing Hannah's scream, we were hoping to mimic a scenario in which Hannah was being threatened by an even more dominant female or male. When this happens under natural conditions, females often "redirect" their frustration by seeking out and threatening a more subordinate individual. (Redirected aggression will be familiar to anyone who has ever picked a quarrel with a spouse, friend, or child after an unpleasant disagreement with a colleague at work.) We predicted that, if grunts really do serve a reconciliatory function, Nimi would be more likely to interpret Hannah's scream as a sign of possible renewed aggression if Hannah had not grunted to her than if she had. She should therefore look in the direction of the speaker for a relatively long duration. But if Hannah had grunted to Nimi after the fight, Nimi should look toward the speaker for only a short duration, if at all. In this case, her response should be no different from what it would be in the absence of any recent aggression.

Note that in this experiment, as in many others, our within-subject design meant that the same subject heard the same individual's scream in all three trials. In this experiment we had, by definition, no control over trial order for two of the conditions. We did, however, vary the order of the third "no prior fight" condition, so that for some subjects this trial was played first, for others second, and for others third.

Like all playback experiments, these were exasperating to conduct. To begin, it took months to build up a library of clearly recorded screams from different females. Further, the design of the experiment required that we play the dominant opponent's scream to the subordinate subject between 10 and 30 minutes after the original fight. As a result, when one of us had witnessed a fight, we had to race frantically through the baboon group to locate the other observer while simultaneously keeping track of the two females to determine whether the dominant grunted to her victim. In 1992 and 1993, before we had two-way radios, these efforts were often ludicrous to behold. Next—while still keeping an eye on the two combatants—one person prepared the video camera as the other searched laboriously through the library of calls on the tape recorder to locate the dominant animal's scream. Again, before the advent of digital audio players this took an agonizingly long time. In a vexing number of cases, we lost track of one of the females before we could conduct a trial. Equally often, the two females simply loitered in the same general vicinity, in full sight of each other, until the allotted time had expired.

Finally, the experimental design demanded that we observe the same two females fight at least twice. After one fight, the dominant female had to grunt to her victim; after the other, she had to remain silent. Some individuals were simply not accommodating. It was easy to observe Sylvia, the second highest-ranking female at the time, threatening someone. Known locally as the "Queen of Mean" and so cantankerous and irritable that she willed herself to live long past the age of 20, Sylvia cut a swathe as she moved through the group, scattering subordinate females and juveniles, and biting or whacking anyone who failed to move out of her way. The problem was that she almost never grunted to her victims. We had the opposite problem with Beth, the elderly third-ranking female. Beth hardly ever threatened anyone, and when she did she almost invariably hugged or grunted to her opponent afterwards. As with Sylvia, we could complete one part of the trial, but not the other.

We were eventually able to complete 15 sets of three trials involving 15 different female-female pairs. Subjects came from several different families and many different ranks. When we examined our results, we found that subjects did indeed respond strongly to their former aggressor's scream if the aggressor had not grunted to them after their fight. Conversely, if the aggressor had grunted, subjects responded as weakly to her scream as they did in the absence of any prior aggression. Apparently, subjects inferred that their former opponent was more likely to renew aggression if she had not grunted to them than if she had. Grunts seemed to serve a reconciliatory function (Cheney et al. 1995b).

We next designed a more ambitious and devious experiment in which we attempted to *mimic* vocal reconciliation. These experiments demanded that we observe, on three separate occasions, two females (say, again, Hannah and Nimi) fight and then separate without grunting or reconciling in any way. In the test trial, we simulated vocal reconciliation by playing Hannah's grunt to Nimi within five minutes after their fight (in 85% of trials we were able to play the grunt within two minutes after the fight). We then followed Nimi for half an hour to observe whether she approached Hannah or interacted with her. Nimi's behavior in the test condition was compared against her behavior in two control conditions. In the first control, we played no grunt to Nimi, but simply followed her for half an hour. This control allowed us to determine whether Nimi's behavior after hearing Hannah's grunt was different from her behavior after she heard no grunt. But we also wanted to guard against the possibility that Nimi would regard *any* dominant female's grunt as a reconciliatory signal. So, in a second control trial we played the grunt of another dominant female unrelated to Hannah (say, Beth) to Nimi after Hannah had threatened Nimi. We were eventually able to complete a full set of three trials on 35 pairs, involving 17 different subordinate subjects.

This experiment was a bit underhanded, because we were fooling the subordinate victim into believing that her opponent had reconciled with her when in fact she had not. But we were not too concerned about the ethics of this pretense. All of the trials were conducted after relatively minor tiffs involving a lunge or brief chase, so we did not expect that the subject would be risking life or limb if she approached her former opponent under the delusion that she had reconciled with her. In fact, subjects were almost never threatened when they approached their former opponent.

Again, and for many of the same reasons, these were maddening experiments to conduct. An additional confound was that we could not force an interaction or compel the females to meet, as we might have been able to do with captive animals. We just had to hope that they would come near each other. In fact, we got lucky. In 74% of trials the subordinate victim and her opponent did come to within two meters of each other in the ensuing half hour. When they did, the subordinates' behavior indicated clearly that they had interpreted the playback of their opponent's grunt as a signal of reconciliation.

After hearing their opponent's grunt, subjects approached their opponent and tolerated their opponent's approaches—by not moving away—at significantly higher rates than they did under baseline conditions, when no aggression had occurred (Cheney and Seyfarth 1997). These results were consistent with our observational data, which indicated that rates of approaching and grunting increase substantially after conflict (Silk et al. 1996). Apparently, a reconciliatory grunt causes subordinate females not just to relax but to seek out their former opponent.

In contrast, if subjects heard either no grunt or the grunt of a different dominant female, they continued to avoid their opponent and retreated from her approaches. Subjects approached their former opponent in only 2% of the follows conducted after playback of either no grunt or the control female's grunt. In contrast, they did so in 42% of the follows conducted after playback of their opponent's grunt.

But subjects did not alter their disposition toward any female whose grunt they heard, because hearing the control female's grunt caused no change in their behavior. After playback of the control female's grunt, subjects did not approach that female at higher rates or attempt to interact with her. In other words, the change in their behavior was specific to their opponent. They appeared to regard their opponent's grunt as directed at themselves, and acted as if they regarded the control female's grunt as irrelevant and directed at someone else (Cheney and Seyfarth 1997). In Chapter 8 we consider what reconciliation reveals about baboons' ability to read the intentions of others.

Kin-mediated reconciliation

Regardless of their dominance rank, group life is essential for all female baboons. Their reproductive success is determined primarily by predation and infanticide, and females can diminish the deleterious effects of these twin selective pressures by establishing and maintaining close bonds with kin, adult males, and other adult females. Reconciliation serves the important function of minimizing and ameliorating the disruptive effects of aggression, restoring tolerance among females, and maintaining group cohesion. In baboons, however, direct reconciliation between opponents and their victims occurs relatively infrequently, after only 10–13% of conflicts (Silk et al. 1996; Wittig et al. 2007a). This low frequency may arise in part because subordinate victims tend to steer clear of their opponents after a dispute in apparent fear that the attack will be renewed.

Interestingly, though, a close relative of the aggressor may often attempt to reconcile with the victim by hugging or grunting to her soon after the conflict ends. (Aficionados of Jane Austen's *Pride and Prejudice* will doubtless recall that when Bingley spurns Jane Bennett, she receives the most devastating—and condescending—"reconciliatory" gestures from Bingley's sister, Caroline.) Among the Okavango baboons, kin-mediated reconciliation occurs roughly twice as often as direct reconciliation by the aggressor herself (Wittig et al. 2007a). If victims treat a friendly gesture or vocalization from their opponent's kin as a proxy for reconciliation with the opponent herself, then reconciliation among female baboons occurs following 30–40% of conflicts, a level similar to that found in chimpanzees, where dominance relations among females are less despotic and nepotistic (reviewed by Wittig and Boesch 2003; see also Aureli and de Waal 2000). In Chapter 6, we describe playback experiments to test this hypothesis.

Rank, social networks, bereavement, and stress

In humans, two classes of causal agents interact to promote stress: physical agents, like smoking, excess alcohol, or a diet high in cholesterol, and psychological agents, like chronic tension at work or a death in the family, that exacerbate the damage to already weakened cardiovascular and immune systems. The interaction between physical and psychological factors is well known (see Sapolsky 2004 for review). Many of these psychological agents can be seen clearly in Michael Marmot's (2004) long-term study of health and longevity in over ten thousand British civil servants. Marmot followed civil servants of all ranks, from less well-educated, blue-collar workers to those at the upper reaches of government service. He found the usual relation between health and physical agents like smoking, alcohol, and diet, but even more striking was the link between health and socioeconomic status (SES): employees at the bottom ranks of the civil service were roughly four times more likely than those at the top to suffer from cardiovascular disease and to die at any given age.

The reason for this had little to do with differences in health care. The subjects in Marmot's study were all well-paid government employees in a country with an established national health care service. And the gradient between SES and health remained even when Marmot controlled for factors like smoking, alcohol, and diet. These results have been replicated in many other countries. Regardless of the level of health care they receive, people on the lower rungs of the socioeconomic scale experience poorer health.

The cause appears to be stress. The lives of people at the bottom of the SES gradient differ from those at the top in three important respects: control, predictability, and social support. Even when they live in relatively affluent societies, poorer people often see themselves as comparatively powerless, with little control over their lives and an inability to predict what will happen next. Furthermore, with some notable exceptions, poor people as a group often have less well-developed networks of social support. They are more likely to live alone, and even when they do live with others their friends and family are often also poor or unhealthy, and unable to offer the kind of social and material support that would buffer someone from a stressful event like the loss of a job, divorce, or the death of a close companion. Low socioeconomic status per se does not cause stress. Instead, stress results from the lack of control and support that are correlated with low status (Marmot 2004).

Like Marmot's civil servants, baboons go about their daily lives in an unconstrained manner. They differ in social status, and these differences are correlated with differences in their access to resources. They also have support networks of varying size and strength. Finally, like civil servants, baboons are subject to a variety of social and environmental stressors—including the loss of rank, predation, and infanticide—that are unpredictable and uncontrollable and have a significant effect on survival and reproduction.

But while unpredictability and the loss of social support are highly stressful to baboons, low-ranking animals do not necessarily experience the most stress. Indeed, numerous studies of nonhuman primates have revealed no clear relation between dominance rank and glucocorticoid levels. Instead, rank instability, the degree to which dominants harass subordinates, and levels of social support appear to play more important roles (reviewed by Abbott et al. 2003; Goymann and Wingfield 2004; Sapolsky 2005). As we mentioned earlier, low-ranking male baboons exhibit higher glucocorticoid levels than high-ranking males when the dominance hierarchy is stable, but during periods of rank instability, when the reproductive control enjoyed by high-ranking males is threatened, it is high-ranking males who experience the most stress.

When captive female macaques are housed in extremely cramped and confined conditions with unfamiliar cage-mates, the females that fall to the bottom of the hierarchy manifest all of the classic symptoms of depression: they huddle by themselves, their glucocorticoid levels skyrocket, their hypothalamic-pituitary-adrenal axis is severely disturbed, they cease to cycle, and they die at high rates (Shively et al. 2005). But the females in these experiments do not at all resemble those living in stable social groups, because under more normal conditions subordinate females show little evidence of rank-related stress or lowered reproduction (e.g., Stavisky et al. 2001; reviewed by Sapolsky 2005). There is no evidence, from either the Okavango or other areas of Africa, that low-ranking female baboons have generally elevated glucocorticoid levels (Weingrill et al. 2004; Beehner et al. 2005; Engh et al. 2006a,b). They do, however, experience stress when their ranks are at risk of changing. As with humans and male baboons, what is stressful to female baboons is not low rank per se, but the unpredictability and loss of control associated with rank instability.

Indeed, compared with both males and captive females living under very unnatural conditions, free-ranging female baboons lead relatively amiable and tranquil lives. While daily life is not entirely egalitarian, the interests of high- and low-ranking females coincide to a considerable degree. In most species of Old World monkeys the correlation between female rank and reproductive success is only weakly positive (reviewed by Silk 2002; Cheney et al. 2004). As a result, high- and low-ranking females are equally likely to have close relatives available for social support (Silk et al. 2003, 2006a,b). High-ranking female baboons can vent frustration on more targets than can low-ranking females, but the level of harassment they impose upon low-ranking females is moderate.

The greatest stressors affecting female baboons are unpredictable events over which they have little control: predator attacks and infanticide. And, as we have discussed, those individuals who are most directly affected by these events—females who have lost a close relative and lactating females whose infants are at risk—experience the greatest stress. In response to these stressors, female baboons take active steps to broaden and strengthen their social networks. Lactating females at risk of infanticide attempt to establish a close friendship with an adult male, and those who succeed in doing so show significantly lower glucocorticoid levels.

The importance of social support during times of stress is particularly evident when a female loses a close relative to predation. In humans, bereavement and feelings of loneliness are associated with increased cortisol production, declines in immune responses and, in some cases, increased mortality (e.g., Irwin et al. 1987; Cacioppo et al. 2000; McCleery et al. 2000; reviewed in Segerstrom and Miller 2004). These effects, however, can be mitigated by social support (Thorsteinsson and James 1999). Social support seems to be particularly important for women's mental health (Taylor et al. 2000; Kendler et al. 2005). Social support is also important for the health and reproduction of nonhuman primates. When captive female macaques are socially isolated or placed into a novel group, they experience increased stress. This stress can be dampened considerably by the presence of a preferred grooming partner (Gust et al. 1994). The same is true of Okavango baboons.

As we described in Chapter 3, female baboons who lose a close relative to predation experience elevated glucocorticoid levels. This increase, though, is only evident in the month following their relative's disappearance; by the second month, glucocorticoid levels have returned to baseline. The relatively temporary effect of a relative's loss on females' stress levels seems to occur because bereaved females attempt to cope with their loss by extending their social network.

Because females concentrate much of their grooming on close kin, we initially predicted that females who had lost a close female relative would show a decrease in the number of grooming partners. Instead, the opposite occurred. When we examined the grooming behavior of bereaved females during the three months immediately following their loss, we found that both their number of grooming partners and their grooming rate increased significantly compared with the months before. In contrast, control females in the same reproductive state showed no such increase (Engh et al. 2006b). Apparently, bereaved females attempted to compensate for the loss of a grooming partner by broadening and strengthening their grooming networks.

These attempts were particularly poignant in the case of Sylvia, the erstwhile "Queen of Mean." By the time that her daughter Sierra was killed by a lion, Sylvia was 21 years old and had lost some of her malevolent edge. When Sierra died, Sylvia seemed to become visibly despondent—Sierra had been her primary grooming partner (Fig. 19).

Figure 19. After Sierra was killed by a lion, Sylvia lost her primary grooming partner. Photograph by Anne Engh.

Soon, however, Sylvia embarked on a grooming campaign. She began to approach other females, uncharacteristically grunting and attempting to groom them. Her particular partner of choice was, oddly, Atchar, a member of the low-ranking Balo matriline. But Sylvia's reputation had evidently preceded her and establishing this new relationship was not easy. Atchar usually fled in panic whenever Sylvia approached giving her rusty, raspy grunts.

From a physiological standpoint, it is not surprising that females attempt to establish bonds with other females after the loss of a companion, and that this increase in social contact has a stress-reducing effect. It is well established that the stress response in both animals and humans can be mitigated by social contact and affiliation (Sapolsky et al. 1997; reviewed by Carter 1998; Panksepp 1998; Aureli et al. 1999). The pituitary hormone oxytocin, which can be released by physical contact (Uvnas-Moberg 1997), seems to play an important role in affiliation-mediated stress reduction, and its effect is particularly pronounced in females (reviewed by Taylor et al. 2000). Oxytocin both inhibits the release of glucocorticoids and promotes affiliative behavior, including not only maternal behavior but also an increased tendency to associate with other females.

The data on bereavement in baboons are especially interesting because previous studies of stress and social isolation in animals have been conducted on group-living animals deprived of all social companions. In contrast, the Okavango baboons who lost a close companion were more similar to humans experiencing loss, because they were not separated from their social group and could still interact with other relatives and companions. Even in the presence of familiar group-mates and other kin, however, these females experienced a stress response, and they apparently sought to alleviate it by broadening and strengthening their social relationships. Similar behavior occurs in the Amboseli baboon population, where females who lose a close relative seek to establish relationships with more distant kin, and those who lack even distant kin seek to establish relationships with unrelated females (Silk et al. 2006a,b).

The stress response is adaptive over the short term in large part because it induces individuals to take active measures to alleviate it. But because it is harmful if maintained over long periods of time, the stress response is also highly selective. Confronted with an array of potential stressors, male and female baboons each exhibit elevated glucocorticoids only in response to those events that directly affect their own survival and reproduction.

In the Okavango, a female's lifetime reproductive success is determined primarily by longevity (which depends largely on predation) and infant survival (which depends largely on infanticide). A female's rank is relatively unimportant in determining her reproductive success, though it may play a crucial role in years of drought or poor floods. Perhaps as a result, females exhibit a stress response primarily to predation and the threat of infanticide, and, to a lesser extent, rank instability. Conversely, females do not show a stress response to several events that do not directly affect them, even though they cause a general uproar in the group at large: instability in the ranks of natal and other resident males; the immigration of a potentially infanticidal male when the female does not have a vulnerable infant; and instability in the female dominance hierarchy if the female's own status is unaffected.

Adult males show a similarly egocentric stress response. A male's lifetime reproductive success is determined primarily by his dominance rank and, to a lesser extent, his longevity. Mirroring these selective pressures, male glucocorticoid levels are most strongly affected by immigration, instability in the male dominance hierarchy, and predation.

Baboons' stress responses also demonstrate that even environmental challenges like predation are inextricably linked with social ones. To survive and reproduce, baboons must not just avoid predation and find sufficient food to support themselves and their infants; they must also create, manage, and maintain the social relationships that buffer and support them in these endeavors. Marais may have been overstating matters when he described the life of a baboon as "one continual nightmare of anxiety." But however purple his prose, one thing is clear: in baboons, as in humans, many causes of stress—and all of its alleviation—are fundamentally social.

SIX

Social Knowledge

He knew all the ramifications of New York's cousinships; and could not only elucidate such complicated questions as that of the connection between the Mingotts (through the Thorleys) with the Dallases of South Carolina, and that of the relationship of the elder branch of Philadelphia Thorleys to the Albany Chiverses..., but could also enumerate the leading characteristics of each family; as, for instance, the fatal tendency of Rushworths to make foolish matches.

EDITH WHARTON, 1920: *THE AGE OF INNOCENCE*

Like the doyens of Edith Wharton's New York society, baboons must master a formidable social calculus if they are to survive and reproduce. Throughout the Okavango Delta, each group of baboons has its own version of the "50 families" that made up New York society, arranged—as Wharton would have appreciated—in a linear dominance rank order based on matrilineal inheritance. In the Okavango as in New York, social ranks among females may be stable for years or even generations, only to erupt suddenly in a challenge to the existing order that pits one extended family against another. The ranks of males who have left their families to live elsewhere seem always in flux.

Cutting across female baboons' largely predictable relations based on rank and kinship are more transient social bonds. Some are formed between low- and high-ranking females from different families who groom and form alliances with each other. Ruby's friendship with the ageing termagant Sylvia, described in Chapter 5, is a good example. Others arise when a male and female form a sexual consortship, or when a female with a young infant forms a friendship with a male for protection against infanticide. These cross-family ties, sexual liaisons, and friendships come and go, like the flings, trysts, and Machiavellian alliances in any soap opera.

What sort of intelligence is required to navigate this social landscape? Baboons certainly *seem* to know a great deal about their companions, but this could just be anthropomorphism on our part. After all, when we see a center fielder move instantly and effortlessly to the spot where a fly ball is about to land, we might be tempted to conclude that he is calculating the instantaneous rate of change in a parabolic function. But that is probably not what is going on in his mind. He executes his catch unthinkingly, without explicit deliberation (McBeath et al. 1995). In a similar manner, baboons' apparently sophisticated knowledge of rank, kinship, and other social relations might be based on a few simple unconscious, implicit rules governed by learned associations. In this chapter we take a closer look at baboons' knowledge of their social companions and examine the mechanisms that underlie it.

Knowledge of other animals' dominance ranks

As we saw in Chapters 4 and 5, linear, transitive dominance relations are a pervasive feature of baboon society. To place herself within this hierarchy, a baboon might simply take note of who is dominant and who is subordinate to herself and leave it at that. Alternatively, she might also distinguish among the relative ranks of others, recognizing, for instance, not only that she is subordinate to Sylvia and dominant to Balo, but also that Sylvia is subordinate to Swallow and Balo dominant to Jeanette. If rank were determined by a physical attribute like size, recognizing other individuals' relative ranks would be easy. Among female baboons, however, there is no relation between rank and size, condition, or age. As a result, the problem is considerably more challenging.

There are hints from their behavior that monkeys do recognize other individuals' relative dominance ranks. Vervet monkey females, for example, solicit grooming from others by presenting a part of their body to them. The solicited individual may or may not accept the invitation. In general, females are most likely to accept the solicitations of the highest-ranking female, next most likely to accept those of the second-ranking female, next most likely to accept those of the third-ranking female, and so on down the line (Seyfarth 1980). Monkeys also seem to be aware of their *own* dominance ranks relative to others. When a

dominant female vervet or baboon approaches two lower-ranking females who are grooming, it is almost invariably the lower-ranking of
the two groomers who moves away (Cheney and Seyfarth 1990, 2005).
In so doing, the lower-ranking female behaves as if she recognizes that
although both she and her grooming partner are lower-ranking than
the approaching female, she is the more subordinate. In turn, the female who remains seated behaves as if she recognizes that she is relatively more dominant than her erstwhile grooming partner. This kind
of evidence is, of course, not definitive. Perhaps the more subordinate
female moves away not because she recognizes her own relative rank
but because she is now surrounded by not one but two females who
outrank her.

Knowledge of other individuals' ranks is also evident in monkeys'
choice of alliance partners. As we have mentioned, female vervets, macaques, and baboons typically support the higher-ranking of the two
opponents when forming alliances with lower-ranking individuals
(Cheney 1983; Chapais 2001; Silk 2002). Similarly, when recruiting alliance partners, monkeys often appear to be assessing not only their own
rank relative to a potential ally but also the rank relation between the
ally and their opponent. In her study of captive male bonnet macaques
(*Macaca radiata*), for example, Silk (1993, 1999) observed that males consistently tried to recruit allies that outranked both themselves and their
opponent. Furthermore, their choice of alliance partner varied depending on the rank of their opponent. If a male was involved in a fight with
the seventh-ranking male, he would attempt to solicit the aid of, say,
the fifth-ranking male. But if his opponent was the 12th-ranking male,
he would attempt to recruit the ninth-ranking one. If the male dominance hierarchy had remained stable, memorizing each male's rank
might not have been a difficult task, but each month roughly half of the
16 males changed rank. The males' behavior suggests that they carefully
monitored all aggressive interactions and constantly updated their list
of ranks.

But these are all observational accounts, open to multiple interpretations. To test baboons' knowledge of rank relations more directly, Joan
Silk and we designed an experiment that took advantage of the fact that
baboon vocalizations are individually distinctive and predictably used
only in certain contexts.

As we have mentioned, high-ranking female baboons often grunt as
they approach mothers and try to handle their infants (Chapter 5). These
grunts seem to facilitate social interactions by appeasing anxious mothers, because lower-ranking mothers are much less likely to move away

Figure 20. Low-ranking females often give fear barks when higher-ranking individuals attempt to handle their infants. Infant handling is usually considerably gentler than this. Photograph by Adrian Bailey/Aurora.

if the approaching female grunts (Cheney et al. 1995b). Occasionally, though, in spite of the higher-ranking female's friendly grunts, a low-ranking mother will utter a fear bark as the female approaches (Fig. 20). Fear barks are unambiguous indicators of subordination: a female never gives them to someone who ranks lower than she. These vocal interactions set the stage for a playback experiment based on what psychologists call a "violation of expectation" paradigm.

Violation of expectation tests first became prominent in studies of cognitive development in human infants. In a typical experiment, a four-month-old sits on her mother's lap, facing a small TV screen. As the baby watches, a solid rectangle appears on the screen with a vertical rod behind it. Parts of the rod stick out above and below the rectangle. The rod oscillates back and forth like a metronome, to the left and right, behind the rectangle. The display continues until the baby begins to look elsewhere, presumably because she has lost interest. Suddenly the display changes and the rectangle is removed. In the control condition, the rod is still there, oriented in the same way and moving back and forth as before. By contrast, in the experimental condition the baby now sees that in fact there were two pieces of rod behind the rectangle that moved back and forth in unison. Babies look for much longer at the broken rod than at the intact rod, presumably because they find it surprising. Their surprise suggests that they had assumed, as any adult would, that the rod behind the rectangle was a single piece. The broken rod thus violated their expectations. Experiments like these reveal that even very young infants have expectations about how physical objects behave: if two identical objects move in unison behind an occluder, they must be connected (e.g., Kellman and Spelke 1983).

The logic behind such experiments is this: If you want to know whether an organism knows *p*, present it with evidence that *p* is true. The organism should not be surprised—that is, it should not react strongly. Then, in a test that is as similar as possible to the first, present it with evidence that *p* is false. Now it should respond with surprise or puzzlement or by seeking additional information, because these new data are at variance with what it thinks to be true.

Applying this logic to baboons, we created call sequences mimicking an interaction that violated the female dominance hierarchy. The sequence consisted of a series of grunts originally recorded from a lower-ranking female (say, the eighth-ranking Shashe) combined with a series of fear barks originally recorded from a higher-ranking female (say, the third-ranking Beth). This sequence violated the female dominance hierarchy because Beth would never give fear barks to the likes of Shashe. As a control, we retained the anomalous sequence, but added the grunts of a third female (say, the alpha female Stroppy) who ranked higher than Beth. This sequence was consistent with the female dominance hierarchy, because it mimicked a scenario in which Stroppy and Shashe were both grunting to Beth and Beth was giving fear barks to Stroppy. In trials separated by several days, we played both sequences to an unrelated female (say, the 17th-ranked Amelia) and filmed her response.

We should reemphasize here that most, if not all, of the vocalizations given by baboons—and indeed other primates, mammals, and birds—are individually distinctive (Hammerschmidt and Todt 1995; Rendall et al. 1996; Owren et al. 1997; Fischer 2004; reviewed in Snowdon 1990; Hauser 1996; Bradbury and Vehrencamp 1998; Ghazanfar and Santos 2004). Thus, when a baboon hears a fear bark, grunt, threat-grunt, scream, bark, or wahoo, she extracts information not only about what is occurring but also about the caller's identity. This information is crucial, because it strongly influences how the listener will respond. Depending upon who is giving a threat-grunt, for example, a listener may decide to enter the dispute and form an alliance, to ignore the interaction, or to beat a hasty retreat.

On the assumption that baboons recognize other individuals' dominance ranks, we predicted that subjects would respond with surprise—

by looking in the direction of the loudspeaker for a longer duration—when they heard the sequence that violated the female dominance hierarchy. In contrast, when they heard the sequence that was consistent with the hierarchy, they should respond with a short ho-hum glance, if they even looked at all. We played the paired sequences to 18 females, using nine different pairs of consistent and inconsistent call sequences. Subjects looked in the direction of the loudspeaker significantly longer when they heard the sequence that suggested a reversal in female dominance ranks (Cheney et al. 1995a). They seemed to recognize who outranked whom and responded more strongly when their expectations were violated.

One might, of course, argue that subjects responded to the anomalous grunt–fear bark sequence because they had never heard that particular combination of call types and signalers before. But the control sequence preserved this anomaly; it simply added another female's calls to make the sequence logical with respect to the female dominance hierarchy. We could also rule out the possibility that call sequences that violated the existing hierarchy were more salient, because the consistent sequences were actually of longer duration than the inconsistent ones. There was also no relation between subjects' responses and the particular signalers involved.

Independent support for the hypothesis that baboons recognize other individuals' dominance ranks comes from experiments carried out with Dawn Kitchen. To test the hypothesis that males recognize the rank relations that exist among other males, individual males were played sequences of wahoos that mimicked a contest between either adjacently ranked or disparately ranked males. To control for the fact that wahoo contests involving adjacently ranked males occur more often than those involving males of disparate ranks (Chapter 4), only the calls of adjacently and disparately ranked males who had interacted at the same rate during the past six months were used in the experiments.

High-ranking male subjects responded significantly more strongly to the playback of a wahoo contest between males of disparate ranks than to the playback of a contest between males of adjacent ranks. This result might have arisen because adjacently ranked males compete with one another in many different contexts, whereas males of disparate ranks tend to compete only when the resource at stake is highly valued—meat, a sexually receptive female, or an infant vulnerable to infanticide. Whatever the explanation, results suggested that males recognize each other's wahoos and can assess the distance in rank between any two males (Kitchen et al. 2005b). The result is particularly striking because, like the male bonnet macaques described earlier, male baboons change ranks often.

Knowledge of other animals' kinship relations

Baboons and other monkeys doubtless recognize their own kin, or close associates. (For ease of discussion, we will use the term "kin" as a proxy for "close associate.") A baboon group, however, is comprised of many adult females and juveniles, each of whom maintains close bonds with her own kin. Can baboons recognize the close bonds that exist among others? Can they take a non-egocentric perspective of their social group and recognize its many different social networks?

Some of the first evidence that monkeys recognize other individuals' social relations emerged as part of a relatively simple playback experiment conducted many years ago to test individual vocal recognition in vervet monkeys. We had noticed that mothers often ran to support their juvenile offspring when the offspring screamed during aggressive interactions or rough play. This observation suggested that mothers recognized their offspring's calls, but we wanted to test the idea experimentally. So we conducted a playback experiment in which we played the distress scream of a juvenile vervet to a group of three adult females, one of whom was the juvenile's mother. As expected, mothers consistently looked toward or approached the loudspeaker for longer durations than did the two control females. But we also found that control females often looked at the mother when they heard the juvenile's call, and they sometimes reacted even before the mother herself had responded. They behaved as if they associated the call with a specific juvenile, and that juvenile with a specific adult female (Cheney and Seyfarth 1980; for another, more elegant experiment on captive macaques, see Dasser 1988).

Years later we tried to replicate this result with baboons. In these experiments, two unrelated adult females served as subjects. By definition, one was dominant to the other. For each pair of subjects, we created three sequences of calls consisting of two other individuals' threat-grunts and screams. This call combination was designed to mimic a common aggressive interaction in which a higher-ranking baboon gives threat-grunts to a lower-ranking animal and the lower-ranking animal screams. In the first control condition, both of the apparent combatants were unrelated to the subjects. In the second, one of the combatants was a close relative of the dominant subject, while the other was unrelated to either female. In the third, test, condition, one of the combatants was a close relative of the dominant subject and the other was a close relative of the subordinate subject.

In conducting these trials, we first ensured that the individuals whose calls were to be played were not in the immediate vicinity (within 50 meters). We then waited until both subjects were seated within roughly seven meters of each other but not interacting, and then played one of the call sequences. We were eventually able to play all three sequences to 26 pairs of females.

When we played the sequence that mimicked a dispute between two individuals unrelated to the subjects, they showed little or no reaction. When we played the sequence that involved a relative of the dominant subject, the subordinate looked at the dominant, as if she recognized that the fight involved that female's relative. The dominant, however, rarely looked at the subordinate (why should she?). But when the squabble involved their relatives, both females looked at each other, as if they were asking, "Hmm. Your relative is fighting with mine. What are we going to do about this?" Equally striking, the dominant subject was more likely to seek out the subordinate subject and supplant her in the half hour that followed these trials than in the half hour that followed the two control sequences. In other words, both subjects behaved as if they recognized that a conflict between their families had occurred, and that this conflict was temporarily disrupting their relationship. Conversely, there was a greater tendency for both the dominant and the subordinate subject to approach each other and interact in a friendly manner following the two control trials than following the test trial (Cheney and Seyfarth 1999).

These experiments provide direct support for the hypothesis that baboons recognize other individuals' kin. They also suggest that females' behavior is influenced not just by their own recent interactions but also by the recent interactions of their relatives. Evidently, just hearing a particular type of interaction is sufficient to change baboons' behavior toward other group members.

Vocal alliances

Natural patterns of aggression also reflect monkeys' knowledge of their group's social networks. As we mentioned in Chapter 5, in many monkey species an individual who has just threatened or been threatened by another animal will often redirect aggression by attacking a third, previously uninvolved, individual. Judge (1982) was the first to note that redirected aggression does not always occur at random. He found that rhesus macaques do not simply threaten the nearest subordinate individual, but that they target a close matrilineal relative of their recent opponent. Similar kin-biased redirected aggression occurs in Japanese macaques (Aureli et al. 1992) and vervets (Cheney and Seyfarth 1986, 1989).

Baboons also show their knowledge of other individuals' kinship relations in their responses to "vocal alliances." As we mentioned earlier, like others before us we had been puzzled by the fact that adult female baboons form alliances at relatively low rates. They do, however, appear to form "vocal alliances," by uttering threat-grunts when they see one female threaten another. Often, the vocal bystander is a close relative of the more dominant female (Wittig et al. 2007b), suggesting that she is supporting her relative, but because she is usually glaring balefully at both of the combatants it is difficult to determine objectively whom she is threatening. Together with Roman Wittig and Cathy Crockford, therefore, we designed another playback experiment to test this hypothesis. Thirteen females served as subjects.

Each subject was played the same higher-ranking female's threat-grunts under three different conditions. In the "kin support" condition, the subject heard the female's threat-grunts within five minutes after she had been threatened by that female's close relative (sister, mother, or daughter). So, for instance, if Comet had threatened Leko, Leko would hear the threat-grunts of Comet's sister, Charade. We compared Leko's response in this condition with her responses in two other conditions. In the "non-kin support" condition, Leko again heard Charade's threat-grunts, but this time after she had been threatened by a dominant female from a different matriline (say, Sylvia). In the control condition, Leko heard Charade's threat-grunts after she had either groomed with or engaged in a friendly interaction with Comet.

On the assumption that bystanders' threat-grunts function as vocal alliances, we predicted that, following an aggressive interaction, subordinate subjects would infer that the threat-grunts were directed at themselves and causally related to the recent dispute. In contrast, in the absence of a recent dispute, we predicted that they would infer that the threat-grunts were directed at someone else. We therefore predicted that subjects would respond more strongly to the call playbacks and avoid the aggressor and her matrilineal relatives for a longer period of time after being threatened than in the absence of a recent fight. If kin and non-kin vocal alliances are equally effective, subjects' responses in the kin support and non-kin support conditions should have been similar. If, however, kin support is more effective than non-kin support, subjects should have responded more strongly to the same female's threat-grunts after being threatened by that female's close relative than after being threatened by a female unrelated to the signaler.

After receiving aggression, subjects responded strongly to the threat-grunts of their aggressor's relative. They looked back and forth toward the speaker for a much longer period than they did in the other two conditions. In addition, in the next hour they avoided both their aggressor and other members of their aggressor's matriline. If they did come into proximity of a member of this matriline, they behaved submissively. In contrast, hearing the threat-grunts of a dominant female unrelated to their aggressor had little effect on subjects' behavior. Similarly, if they had not recently been threatened, subjects ignored the threat-grunts and did not try to avoid the signaler or her relatives (Wittig et al. 2007b).

When inferring whether a threat-grunt is directed at themselves or at someone else, therefore, subordinate female baboons appear to take into account the signaler's identity, her relationship with her opponent, and the nature of recent interactions. If they have recently been threatened by a more dominant female, they treat the threat-grunts from that female's close relative as an additional threat against themselves. As a result, the threat-grunts of kin function as vocal alliances. In contrast, if they have not recently been threatened, or if they have been threatened by a female unrelated to the signaler, they seem to assume that the threat-grunts are directed at someone else (Fig. 21). We take up the interesting issue of baboons' inferences about the intended target of a vocalization in Chapter 8.

Kin-mediated reconciliation

Female baboons often grunt to the victims of their relatives' aggression, as if acting as proxies for their relatives. Indeed, kin-mediated reconciliation occurs at double the rate of direct reconciliation (Chapter 5). Kin-mediated reconciliation might substitute for direct reconciliation when aggressors are not motivated to initiate friendly contact or when victims avoid their aggressor's approaches. In chimpanzees, postconflict friendly behavior by an uninvolved bystander toward the victim of aggression has operationally been termed "consolation" (Wittig and Boesch 2003) and has in some cases been taken as evidence for empathy (Preston and de Waal 2002; see also de Waal and Aureli 1996). The apparent absence of consolation in monkeys has, in turn, been

Figure 21. In large social groups, it can be difficult for individuals to determine when a grunt is directed at them as opposed to someone else. Photograph by Adrian Bailey/Aurora.

interpreted as evidence that monkeys are unable to empathize because they cannot attribute mental states like fear or anxiety to others. This explanation is controversial, however, in part because there is no clear evidence that even chimpanzees are able to attribute mental states different from their own to others. We discuss this somewhat contentious issue in Chapter 8.

Before debating the existence—or lack—of empathy and consolation in monkeys, it is first essential to demonstrate that the victims of aggression do in fact accept a friendly overture by the relative of their aggressor as a substitute for reconciliation by the aggressor herself. To examine this question, we designed another playback experiment.

In this experiment, females who had recently been threatened by a more dominant female heard the grunt of one of their aggressor's close female relatives (the "reconciling relative," for ease of discussion), to mimic kin-mediated reconciliation (Wittig et al. 2007a). In the control condition, they heard the grunt of a dominant female from another matriline. On the assumption that subjects would treat the reconciling relative's grunt as a proxy for direct reconciliation with their aggressor, we predicted that they would be more likely to approach their aggressor, tolerate her approaches, and interact in a friendly manner with her after hearing the reconciling relative's grunt than after hearing the grunt of the unrelated female. Again, all trials were conducted within five minutes of the dispute, after the subject had separated from her aggressor and was out of sight of the aggressor and all of her relatives.

Subjects responded as if they assumed that the reconciling relative's grunt was directed at them and was causally related to the recent fight. Upon hearing the grunt of their aggressor's close relative, female baboons looked toward the speaker more often and for a longer duration than upon hearing the grunt of a female unrelated to their aggressor. Furthermore, in the hour following playback of the reconciling relative's grunt, subjects' latency to tolerate their aggressor's proximity was significantly shorter, and their first interaction with the aggressor was significantly less likely to be submissive. Their disposition toward the reconciling relative was also affected. After hearing this female's grunt, subjects' first interaction with her was less likely to be submissive than if they had heard the grunt of another dominant female. Subjects either approached or tolerated the approach of their aggressor or the reconciling relative in 12 (75%) test trials, compared with only one (6%) control trial. In four of these test trials, they also engaged in friendly behavior, such as embracing or grooming, with the aggressor.

Interestingly, although the grunt of the reconciling relative functioned to reconcile subjects with their aggressor, they did not generalize their response toward all members of those two females' matriline. Subjects were not more likely to tolerate the proximity of other members of the aggressor's matriline after hearing the reconciling relative's grunt than after hearing the grunt of the unrelated female.

Similarly, as in our other experiments on direct reconciliation (Chapter 5), subjects did not simply alter their disposition toward *any* individual whose call they heard, because their behavior toward the control female was unaffected by that female's friendly grunts. They were not more likely to approach the control female, nor were they more likely to tolerate her approaches. Perhaps because they had not recently interacted with the control female, subjects seemed to interpret that female's grunt as irrelevant to the recent fight and directed at someone else (again, see Chapter 8 for further discussion of this question). In contrast, even though they had also not recently interacted with the reconciling relative, they treated this female's grunt as relevant to the dispute. Apparently, the relative's close bond with the aggressor was sufficient to cause subjects to infer that the grunt must be directed at them (Wittig et al. 2007a).

Call type was also important. Although the reconciling relative's grunt caused subjects to tolerate their aggressor's proximity, hearing the same relative's threat-grunts caused them to avoid her. Subjects were, therefore, sensitive not only to individual identity when assessing a call, but also to the call's function in social interactions.

The experiments on baboons' responses to vocal alliances and kin-mediated reconciliation provide further evidence that baboons recognize other females' kinship relations. However, although they recognize that close kin can serve as proxies for each other, they nonetheless distinguish among the different members of a matriline. Hearing a "reconciliatory" grunt from an opponent's relative changes females' disposition toward the opponent and that relative, but less so toward other members of the opponent's matriline. Baboons do not treat all the members of a matriline as equivalent.

In sum, female baboons recognize other individuals' kin (or close associates), and they modify their interactions with other individuals according to recent events involving those individuals' kin. They view their social group not just as a collection of individuals but as a network of social relationships in which each individual is closely linked to several others. Baboon social relationships are not, as some have suggested, mere anthropomorphisms (Barrett and Henzi 2002, 2005): they exist

Figure 22. Brothers, like sisters, share close social bonds. Two brothers play in a rain puddle. Photograph by Anne Engh.

in the minds of the animals themselves. The information that baboons acquire about their companions—and that underlies their responses to our experiments—can be obtained only by observing the social interactions that occur among others and making the appropriate deductions. Clearly, baboons are skilled at navigating the social landscape.

Although several studies have now suggested that monkeys recognize kinship relations among other group members, we still know little about how they do this. Monkeys might have a concept of "mother" that is distinct from "sister." Or they may simply evaluate others' relationships based on rates of interaction. Because animals interact at high rates with close matrilineal relatives, this rule of thumb could allow monkeys to recognize other individuals' "kin" (Fig. 22). There is currently little evidence that monkeys discriminate kin from unrelated animals that interact at high rates, or that they recognize other individuals' kin by physical traits such as odor or appearance.

More transient social relations

Rank relations among baboon matrilines can remain stable for decades. Male dominance ranks, in contrast, change often. As we have seen, there is good evidence that males can track rank changes among others, and that they update their list regularly, placing themselves and others accurately in the new list.

Additional evidence that baboons actively monitor short-term changes in even very temporary relationships comes from data on males' tracking of sexual consortships. As we mentioned in Chapter 4, most sexual consortships—particularly those that occur at the height of the female's cycle—are formed by the group's alpha male. Unlike male baboons in East Africa, males in the Okavango do not form coalitions to challenge a consort, so as long as the alpha male is with his female there is little opportunity for other males to mate. This does not mean, however, that other males acquiesce passively to this unhappy state of affairs. Instead, they keep a watchful eye on the consorting pair. If the male wanders away from his female, they dash in to attempt a quick mating. And as soon as the alpha male abandons the female after the consortship ends, other males vie to mate with her or to form another consortship.

Cathy Crockford and Roman Wittig conducted a playback experiment to examine how closely males monitor other males' consort status, using as their stimuli male grunts and female copulation calls (Crockford et al. 2007). Males often, but not always, grunt when they approach a female to mate with her, and females almost invariably give loud copulation calls after mating. The subjects in this experiment were nine adolescent and lower-ranking adult males, hopeful suitors all. Each subject appeared in three separate trials: two while a high-ranking male was involved in a sexual consortship, and the third as soon as possible—ideally within several hours—after the consortship had ended.

Playbacks were conducted when the subject was resting or feeding on a sausage fruit or palm nut that would demand at least a few seconds' repose. One loudspeaker was then placed approximately 20 meters to the subject's left and another the same distance to the subject's right. In the test trial, the subject heard the grunt of a consorting male (say, Nat) played from one speaker, and then, several seconds later, his consort's copulation call played from the other. This sequence suggested that the male and his consort had separated and that the female was "sneakily" mating with another male.

The second trial was similar to the first, except that the subject heard the grunts of a different, nonconsorting high-ranking male (say, Fat Tony) played from the first speaker and the same female's copulation call played from the second. This sequence implied that the consortship was still ongoing, but that another male was in the general vicinity—a very common occurrence.

Finally, in the third trial, conducted as soon as possible (and always within 24 hours) after the consortship had ended, the same subject again heard the former consorting male's (Nat's) grunts and the same female's copulation call played from different loudspeakers. As in the first trial, this sequence suggested that the male had separated from his female and that she was mating with another male. After consortships have ended, this is not at all unusual.

If males constantly monitor the status of other males' consortships, they should have responded most strongly to the test trial. The second trial should have been unsurprising, because it provided no new information about the consortship's status. The third should also have been unsurprising, but only if the subject was keeping a close, even obsessive, eye on the consortship's status.

This is precisely how subjects behaved. When they heard the two calls coming from different speakers when the consortship was still ongoing, they looked sharply toward the speaker that had played the copulation call, and in 67% of trials rapidly approached it. Their behavior suggested that they inferred that the consort pair had temporarily separated, that the female was engaged in a sneaky mating, and that further mating opportunities might be possible.

In contrast, subjects responded much more weakly when they heard a nonconsorting male's grunts and the female's copulation call played from different speakers. Their relative apathy suggested that they interpreted this call sequence as indicating that the consortship was still active, and that a nonconsorting male was simply nearby. This information did not violate their expectations, nor did it signal a mating opportunity.

Subjects' responses were equally nonchalant when they heard the former consort male's grunts and the female's copulation call played from separate speakers after the consortship had ended, even when it had ended only a few hours ago. Their weak responses could not be explained by a decrease in the attractiveness of the female, because in each case the female continued to mate with lower-ranking males for several days after the playback. Instead, the males appeared to respond weakly because the call sequence provided redundant information: they already knew that the consortship had ended. Apparently, male baboons are such assiduous voyeurs that they can deduce instantly, on the basis of the spatial and temporal juxtaposition of two calls, that a female has momentarily separated from her consort (Crockford et al. 2007).

The ability to monitor, or eavesdrop upon, the interactions of others is thought to be adaptive because it permits animals to assess the strength of other individuals' allies, pair bonds, and competitive abilities without engaging in potentially costly interactions. As we will see in the next chapter, experiments have now demonstrated that a variety of animals are able to recognize other individuals' close associates and dominance ranks. And this knowledge can be acquired rapidly. Chickadees (*Poecile atricapilla*), for example, can determine the relative ranks of two rival males after as few as six minutes' exposure to their singing contest (Mennill et al. 2002). It remains unclear, however, whether any animals other than nonhuman primates are able to keep track of temporary changes in highly transient relationships. Dominance ranks and kinship relations are relatively static attributes of individuals. Consortships, in contrast, are temporary and unpredictable: individual males and females are not always involved in consortships, and their consort status can change over very short periods of time. Baboons, however, are able to monitor the status not only of relatively long-term kinship and dominance relationships but also of very transient sexual relationships.

Evidence that baboons monitor very transient social relationships reminds us that baboons do not simply memorize a few facts about each individual's rank and family membership and leave it at that. Instead, faced with recurring social instability against a background of kinship and rank, baboons maintain a running account of each individual's social relationships and status. The cognitive demands of this job would seem to be much easier if baboons could organize their knowledge hierarchically and group other individuals into distinct categories.

Structured, hierarchical knowledge

Humans do not just recognize the relationships that exist among others; we also classify individuals according to multiple attributes simultaneously. Consider the characters in Shakespeare's *Romeo and Juliet*. Romeo's personality derives from his particular individual attributes—he's a romantic, self-confident male—combined with his membership in a rich, aristocratic family, the Montagues. Juliet's personality is, likewise, an amalgam of her individual traits, her status as an unmarried teenager, and the inescapable fact that she is a Capulet. As the play unfolds and we learn about each character's idiosyncrasies and family allegiances, we form expectations about the kind of social relationships that are likely to result when any two characters come together. When we see Mercutio teasing his friend Romeo in Act III, we dismiss the teasing as trivial because the two are allied with the house of Montague. But when Mercutio later aims his taunts at Tybalt, we regard his behavior as more ominous because Tybalt is a Capulet and the dispute could easily escalate into a bloody confrontation engulfing all the members of both families. Our responses are guided in part by our tendency to organize social relationships into a hierarchical structure—in this case, familial affiliation—that is governed by a functional set of rules: quarrels between families are potentially much more destructive than quarrels within families.

Do baboons make classifications of this sort and form similar expectations about other individuals' behavior? Do they recognize, for example, that the individuals in their group can be classified simultaneously according to both their dominance rank and their membership in a particular matrilineal family? And, if they do, can they recognize the significance of a rank reversal between two females from different matrilines? To examine these questions, Thore Bergman, Jacinta Beehner, and we designed another experiment based on the violation of expectation paradigm.

In separate trials, 19 adult females heard three different sequences of threat-grunts and screams, each mimicking a fight between two females belonging to a matriline different from their own. One sequence of

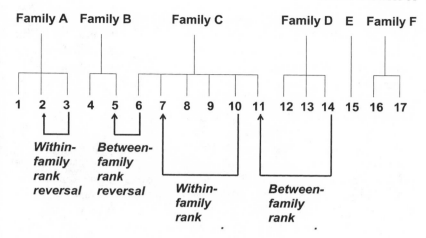

Figure 23. Examples of simulated within- and between-family rank reversals. In a typical between-family rank reversal sequence, for example, a subject might hear a member of family D giving threat-grunts while a member of family C screamed.

calls mimicked a rank reversal between two members of the same matriline (Fig. 23). A typical sequence might consist of the threat-grunts of the 13th-ranked Champagne combined with the screams of her tenth-ranked aunt, Hannah. The second sequence of calls mimicked a rank reversal between two members of different matrilines. In this case, the sequence might consist of Champagne's threat-grunts and the screams of the sixth-ranked Luxe. The third, control sequence of calls mimicked a fight that was consistent with the female dominance hierarchy. In some control sequences the signalers were from the same matriline (Hannah threatens Champagne); in others, they were unrelated (Luxe threatens Champagne). Luckily, we were able to control for rank distance effects in these experiments, because the group contained several large matrilines. As a result, some sequences involving members of the same matriline consisted of individuals who were separated by as many as five positions in rank. Conversely, some of the between-family rank sequences involved females of adjacent rank.

To understand the logic behind these experiments, it is important to remember that, for the most part, female dominance ranks are remarkably stable over time, and that ranks within matrilines are as stable as those between matrilines. Absolute ranks change whenever a female matures or dies, but the relative ranks among females rarely alter. Furthermore, rates of aggression within families are similar to those between families. However, when within-family rank reversals do occur, they typically involve only a minor social adjustment in the rank order among sisters or mothers and daughters and have little effect on rank relationships outside the matriline. In contrast, when a female from a lower-ranking matriline defeats a female from a higher-ranking one, as occurred in 2003, many members of multiple matrilines may lose or gain rank (Chapter 5; see also Samuels et al. 1987; Chapais 1988; Cheney and Seyfarth 1990). Between-family rank reversals are therefore much more momentous than within-family ones.

Because rank reversals involving the members of different matrilines have the potential to influence the ranks of many individuals, we predicted that subjects would respond more strongly to sequences that simulated a between-family rank reversal than to those that simulated a within-family rank reversal. This is exactly what we found. Not surprisingly, females also responded more strongly to between-family rank reversals than to control sequences (Bergman et al. 2003). Subjects' strong response to the between-family rank reversal sequences were not due solely to the anomaly of these calls, because between-family and within-family rank reversals are equally rare. Nonetheless, subjects responded much more strongly to the former.

These experiments provide some of the first evidence that monkeys classify others simultaneously according to both their individual attributes, like rank, and their membership in higher-order groups, like matrilines, and that they do so in the absence of human training. Baboons appear to understand that their group's female dominance hierarchy can be subdivided into familial units, and they seem to make a sharp distinction between rank reversals within and between these units. As a result, they appear to recognize that, although predictable rank relationships are maintained both within and between matrilines, the latter are qualitatively different from the former.

In their recruitment of alliance partners, too, monkeys show evidence of classifying others according to both rank and maternal kinship. Schino and his colleagues (2006) have found that Japanese macaques preferentially attempt to recruit coalition partners who are both higher-ranking and unrelated to their opponent. These observations suggest that Japanese macaques, like baboons, simultaneously take into account information about other individuals' ranks and familial associations when selecting the optimum alliance partner.

Monkeys' knowledge of social structure is probably not explicit—they cannot take a stick and draw kinship diagrams in the sand, any more than a four month-old human infant can explain why she assumes that two pieces of rod moving in unison behind a rectangle must be parts of the same rod. Nonetheless, monkeys combine their knowledge of rank and kinship to perceive that ostensibly similar events can have very different consequences, some affecting the lives of only a few individuals and others affecting the lives of many.

Reaction time, causal inferences, and the perception of intent

Three aspects of the experiments described in this chapter deserve additional mention. First, baboons' latency to respond to playback sequences is often very short. Given the amount of information that must be processed, the speed of their reactions is striking. To recognize that the call sequence "Shashe grunts and Beth fear-barks" violates the existing dominance hierarchy, the listener must recognize the type of call given, consider the rules that govern the use of each call type (fear barks are given only to higher-ranking females, while grunts may be directed to either higher-ranking or lower-ranking females), identify both signalers, and recall their respective dominance ranks. And to recognize that the sequence "Champagne threat-grunts and Luxe screams" is more portentous than "Champagne threat-grunts and Hannah screams"—even though both sequences violate the dominance hierarchy—the listener must add matrilineal membership to her calculations. Baboons make these complicated calculations very quickly, and probably unconsciously.

Second, baboons' responses suggest that they assume a causal relation between vocalizations that are closely spaced in time and location. If a baboon who heard the sequence "Champagne threat-grunts and Luxe screams" thought that these calls were juxtaposed purely by chance, there would be no reason to respond strongly. The sequence only merits a strong response if the listener assumes that the calls occur together because Champagne's threat-grunts *caused* Luxe to scream. Baboons apply these assessments even to indirect causal relations. If a female has recently been threatened, she assumes that a "reconciliatory" grunt from her aggressor's relative is directly related to the fight.

Third, baboons' responses suggest that they make inferences about both the intended target of a signaler's call and the signaler's motivation toward her. When a listener responds strongly to the sequence "Champagne threat-grunts and Luxe screams," she acts as if she assumes that Champagne is intending to threaten Luxe. Although this seems unexceptional for humans, researchers have only recently begun to entertain the possibility that baboons and other monkeys might be able to attribute simple mental states like intent to others. We look more closely at this problem in Chapters 8 and 9.

Social skills are adaptive

If natural selection has led to the evolution of social skills, we should expect to find evidence that these skills increase individuals' reproductive success. As we discussed in Chapter 5, evidence in support of this hypothesis is beginning to emerge from long-term studies of baboons.

In the Okavango, the reproductive success of female baboons depends primarily on their ability to avoid infanticide and predation. Faced with the threat posed by a potentially infanticidal male, females with infants form friendships with males. When they lose a close relative to predation, they seek out new grooming partners and extend their social network (Fig. 24). In Amboseli, females strengthen their relationships with their sisters after their mothers die and strengthen their relationships with more distantly related kin or non-kin if they lose their close relatives (Silk et al. 2006a,b).

Skills in forming relationships appear to enhance a female's lifetime reproductive success. In the longest-running study of baboons to date, Joan Silk, Susan Alberts, and Jeanne Altmann examined the correlates of lifetime reproductive success among 118 females in five different groups living in Amboseli National Park, Kenya. They found that the females with the highest rates of infant survival were not the highest-ranking females but those who were most socially integrated (Silk et al. 2003). Close social bonds, therefore, are highly adaptive. But females cannot maintain close social bonds with an unlimited number of partners. In Amboseli, the number of females' close social bonds seems to plateau at six (mean 1.6, range 0–6) (Silk et al. 2006a,b). This suggests either that females do not have enough time to maintain close relationships with many other individuals or that they simply do not need a larger number of partners to meet their social needs. In order to establish new social bonds after the death of a close partner, however, a female may need to recognize which females already have extensive social networks and which do not, and which males already have close friendships with lactating females. And, in choosing her allies and enemies, a female may need to recognize other individuals' relative ranks and kin relations. How do they do it?

Figure 24. Sirius, the juvenile daughter of Selo, hugs her younger brother, Spud. Photograph by Keena Seyfarth.

Underlying mechanisms

Can a few simple rules explain the complexity of baboons' social knowledge? Some learning psychologists believe that they can, and have argued that monkeys' apparent recognition of other individuals' kin (or close associates) is simply an example of associative learning and conditioning.

It has long been known that laboratory animals like rats and pigeons can be taught to group even very different looking stimuli together if

111

they are all associated with the same reward or outcome. In one series of tests, Schusterman and Kastak (1993, 1998) taught Rio, a California sea lion (*Zalophus californianus*), to group arbitrary symbols into "equivalence classes." Each group consisted of three cards depicting a symbol: for example, a pipe (A_1), a fish (A_2), and a star (A_3). The experimenters arranged the symbols into equivalence classes by displaying one group of cards (say, A_1, A_2, and A_3) next to each other on one side of Rio's enclosure and another group of cards (say, B_1, B_2, and B_3) next to each other on the other side. After a few days' exposure, Rio was presented with one card from the A class and one card from the B class (A_1 and B_1). As soon as she prodded one of the cards with her nose, she was rewarded with food. Assuming Rio chose A_1 rather than B_1, she then received repeated presentations of the same cards, with A_1 always rewarded and B_1 not rewarded, until she achieved a 90% success rate in a block of 10 consecutive trials. Then Rio was tested, first with symbols A_2 and B_2 (transfer test 1) and next with symbols A_3 and B_3 (transfer test 2), to determine whether she had begun to treat all A stimuli as equivalent to each other and all B stimuli as equivalent to each other, at least insofar as they followed the rule "if $A_1 > B_1$ then $A_n > B_n$." Rio performed correctly on 28 of 30 transfer tests.

Schusterman and Kastak argue that these relatively simple equivalence judgments constitute a general learning process that underlies much of the social behavior of animals, including the recognition of social relationships by monkeys. They propose that "features of the environment can become related through behavioral contingencies, becoming mutually substitutable even when sharing few or no perceptual similarities" (1998:1088; see also Dube et al. 1993; Heyes 1994; Wasserman and Astley 1994; Thompson 1995). In other words, even if two baboons from the same family look very different, they can become linked in an equivalence class because they share a history of common association. As a result, when Sylvia redirects aggression against Balo after fighting with another member of the Balo matriline, she does so because all the members of that matriline have effectively become, in Sylvia's mind, interchangeable. And once equivalence classes have been formed, members of the same matriline not only become mutually substitutable but also exhibit transitivity: if $A_1 > B_1$, then $A_2 > B_2$.

This argument has much validity. It is hard to imagine how a monkey could learn that two other individuals were members of the same matriline except by grouping them together by virtue of their high rates of association. When forming alliances, monkeys avoid recruiting a close relative of an opponent precisely because they recognize that the two belong to the same kin group. They redirect aggression against an opponent's relative for the same reason. To a large degree, monkeys do place matrilineal kin into the same equivalence class. At the same time, however, the "equivalence classes" that make up nonhuman primate groups exhibit complexities not present in laboratory experiments.

First, consider the magnitude of the problem. The sea lion Rio was confronted with a total of 180 dyadic comparisons. This is roughly equivalent to the number of different dyads that confront a monkey in a group of 18 individuals. But the number of possible dyads increases rapidly as group size increases. Baboons often live in groups of up to 80 individuals, which contain 3,160 different dyads. The number of triadic combinations is much larger still—a group of 80 baboons is composed of 82,160 different triadic combinations. As a result, monkeys face problems in learning and memory that are not just quantitatively, but also qualitatively, different from those presented in a typical laboratory experiment. This is important, because it is large numbers that may force primates to develop rules to classify their group-mates.

Second, in monkey groups no single metric specifies the associations between individuals. It is, of course, a truism that monkeys can learn which other individuals share a close social relationship by watching them interact. Matrilineal kin, for example, almost always associate at higher rates than non-kin. But no single behavioral measure or rule is either necessary or sufficient to recognize such associations. Aggression often occurs at the same rate within and between families, and different family members may groom and associate with each other at widely different rates. A human observer who quantifies rates of grooming and uses these data to identify all of a female's relatives is apt to make more mistakes than the monkeys do.

Similarly, although monkeys doubtless use high rates of association as one criterion for identifying a close social bond, they seem to recognize that different types of relationships are characterized by different patterns of spatial proximity. As a result, they are not surprised when a playback experiment suggests that a female is not in close proximity to her adult or juvenile daughter. They do, however, respond strongly to playbacks suggesting even a temporary separation between a male and his consort female. They apparently understand that some close social relationships are characterized by continuous spatial proximity, while others are not. They also appear to recognize that some close relationships can be very transient, because the same sequence of grunts and copulation calls that elicits a strong response from subjects when a consortship is still ongoing elicits a much weaker response within hours after it has ended. To recognize that some relationships are defined by close spatial proximity and others not, and that some relationships are enduring and others not, a monkey must take note of a variety of different patterns of aggression, reconciliation, grooming, and proximity. There is no threshold or simple defining criterion for a "close" social bond.

By contrast, the equivalence classes in Schusterman and Kastak's experiments were established by repeatedly presenting a group of three symbols to Rio at the same place and time. This spatial and temporal juxtaposition provided an easy, one-dimensional method for the formation of equivalence classes. As yet, we do not know if monkeys distinguish between matrilineal kin bonds and the equally strong bonds that may form between unrelated females who interact at high rates. If monkeys do make such distinctions, this would provide further evidence that they assess and compare social relationships using a metric based on more than just patterns of association.

Third, class members in monkey groups are sometimes mutually substitutable and sometimes not. In both baboons and vervets, females who hear a juvenile scream will often look toward the juvenile's mother. Schusterman and Kastak (1998) argue that this is because the scream, the juvenile, and the juvenile's mother form a three-member equivalence class in which any one of the stimuli can be substituted for another. But in fact the call, the juvenile, and the mother are not interchangeable in this manner. A female who has a close bond with the juvenile's mother, for example, may interact very little with the juvenile himself. The call is linked primarily to the juvenile and only secondarily to the mother. Indeed, playback experiments on rhesus macaques have shown that, although monkeys do group calls given by members of the same matriline into the same category, they also distinguish among the calls given by different individuals within the same matriline (Rendall et al. 1996). Further, experiments on kin-mediated reconciliation indicate that hearing the "reconciliatory" grunt of their opponent's relative changes females' disposition toward their opponent and that relative but not toward all members of their opponent's matriline (Wittig et al. 2007a). Thus, while female baboons recognize other females' memberships in particular matrilines, they nonetheless treat these members as distinct, not mutually substitutable, individuals.

Fourth, some social relationships are transitive but others are not. In baboon society, if an infant and a juvenile both associate at high rates with the same adult female, it is usually correct to infer that the two are siblings and will also associate at high rates. Similarly, if the female associates at a high rate with a particular adult male friend, it is probably cor-

Figure 25. Males are very protective of their female friends' infants. Betelgeuse grunts to Lizzie's infant, Zoe. Photograph by Anne Engh.

rect to assume that the male is also closely allied to the infant (Fig. 25). It would be incorrect, though, to make the same assumption about the juvenile, because males seldom interact at high rates with their friends' older offspring. In fact, the juvenile is more likely to associate with a different male—the male who was his mother's friend when he was an infant. Similarly, female members of the same matriline often form friendships with different males, and, conversely, the same male may form simultaneous friendships with females from two different matrilines. A close bond between Selo and her daughter Palm, and Palm and her friend Elvis does not imply that Selo and Elvis are also closely bonded. Close bonds between Elvis and his friends Palm and Jackalberry does not predict a close bond between the two females. In fact, their relationship is more likely to be competitive than friendly (Palombit et al. 2001).

Further complicating matters, a baboon can belong to many different classes simultaneously. An adult female, for instance, belongs to a matrilineal kin group, associates with one or more adult males, holds a particular dominance rank, and may be weakly or strongly linked to other females outside her matriline. Here again, the natural situation is considerably more complex than that in most laboratory settings.

Finally, some types of class membership change often. While the rank and kin relations among female baboons are relatively stable, other social relationships change often and unpredictably. Some grooming interactions among unrelated females wax and wane over short periods of time. A female with a new infant receives a lot of attention from unrelated females, and for several weeks she may be groomed more often by nonrelatives hoping to get a chance to handle her infant than by her relatives. Sexual consortships and male dominance ranks fluctuate over periods of days or weeks. Over still longer time scales, females' relationships with other animals change when they give birth and form friendships. Furthermore, while dominance ranks and kinship relations are relatively static attributes of individuals, other relationships are less predictable and highly transitory. Individual males and females are not always involved in a friendship relation or consortship, and their consort status can change from one minute to the next. Baboons, though, are able to monitor the status of these transient, nonobligatory relationships.

In sum, there is no doubt that associative processes and contingency-based learning provide powerful and often accurate means for animals to assess the relationships that exist among others. However, in order to conclude that all primate social knowledge results from simple learning mechanisms, we need proof that these mechanisms can account for behavior as complex as that which occurs in free-ranging primate groups.

Suppose that the sea lion Rio were trained with an array of 80 items (the approximate size of a baboon group), each of which associated at varying rates with all 79 other items but at high rates with a subset of the items. Item A, for example, might associate at a high rate with items B, C, D, and E. B might also associate with these items, but at a different rate than A. B would also associate with some items with which A rarely associated. To complicate matters further, there would also be brief, transient associations of varying durations between pairs of items that cut across the links formed between items that associated at a high rate. Under these circumstances, could Rio, without any training or reward, learn to group the items that associate at high rates into "kin" classes while simultaneously keeping track of the transient pairings that cut across classes? It is entirely possible that Rio would rise to the occasion. In the absence of such an experiment, though, it is impossible to calibrate Rio's and the baboons' respective performances.

Laboratory experiments designed to explain complex behavior using the simplest explanation possible have limited external validity if they leave out the very complexity they hope to explain, or if they depend on extensive training and reinforcement. Baboons, after all, do not derive immediate and predictable rewards from their knowledge of other individuals' social relationships, unless we assume that—like readers of Edith Wharton—they find social spectatorship inherently rewarding. But if we follow this line of argument and assume that baboons are motivated by the inherent value of acquiring gossip, our concepts of reward and reinforcement must become considerably broader and more open-ended than in most laboratory studies of learning.

Social intelligence and the origins of baboon metaphysics

In small groups, individuals can rely on simple associative mechanisms to solve their social problems. A monkey in a group of 10 or even 20 animals need only remember who associates with whom, and who has been nice (or nasty) to her in the past, to predict successfully what others will do and form the most useful social relations. As group size increases, however, the number of dyadic and triadic relations increases explosively, and the simplest forms of associative learning rapidly become inadequate. Impressive though her curmudgeonly memory may be, it may simply not be possible for Sylvia to remember the rank relations in all 82,000 triads in her group—or even the rank relations in all 1,027 triadic relations that involve her.

To solve the problem of large numbers, natural selection may have favored animals that are predisposed to search for ways to arrange their companions into rule-governed classes. For baboons, these classes are based largely on matrilineal kinship and a transitive rank hierarchy. The formation of such classes is an adaptive strategy because it reduces memory load and allows the individual to make quick, accurate judgments of social relationships and predict other individuals' behavior. Put somewhat differently, natural selection has favored individuals who develop theories of social life.

Faced with the problem of remembering long strings of letters, words, or numbers, human subjects learn the string faster and remember it better if some kind of rule allows them to group the items into "chunks." The sequence 12342345345145125123124 is difficult to remember until you see the pattern. The same holds for the sequence 149162536496481100, until you realize that it is the squares of the integers from one to ten. Chunking in humans is adaptive because it increases the capacity of short-term memory (Miller 1956; Simon 1974). In the 14916 ... sequence, you need to remember only one rule, not the sequence of digits. The tendency to chunk is so pervasive that human subjects will work to discover an underlying rule even when the experimenter has—perversely—made sure there is none (Tulving 1962; Macuda and Roberts 1995). Humans are naturally inclined to search for a higher-order rule or pattern that makes the task easier.

Baboons may be the same. Consider the predicament of a young immigrant male whose reproductive success depends almost entirely on his ability to rise rapidly in the male hierarchy. His likelihood of success will depend not just on his fighting skills but also on his ability to learn as much as possible as quickly as possible about relations among his male rivals, the bonds among females, and current male-female friendships and sexual consortships. He could, of course, learn all of these things by painstaking observation: by watching, for hours on end, what each individual does with every other and committing it all to memory. But his task would be eased considerably if he could tackle the problem armed with theories—about matrilineal kinship, transitive rank relations, kin-based alliances, the different causes of dominance rank in males and females, sexual consortships, and friendships. In *The Age of Innocence,* the exotic European Ellen Olenska arrives in New York knowing no one, yet she captivates the community because she learns fast, and she learns fast because she knows the rules. Her social theories are adaptive.

Again, when we say that baboons have social theories we do not mean that they have fully conscious, well-worked-out theories that they can describe explicitly. Nor do we mean that their interactions with others are motivated by their projections, long into the future, of the benefits to be derived from a particular relationship (cf. Barrett et al. 2002). Instead, baboons appear to have implicit expectations about how individuals will interact with one another. Through processes we do not yet understand, they observe the associations among other group members and generate expectations—for example about within- and between-family dominance relations. These expectations constitute a theory because they do not depend on specific individuals, but have general applicability. A baboon's social theories allow her to predict another's behavior even if she has never observed or interacted with that animal. When she meets, say, a new infant who has just been born into a high-ranking matriline, her theories about rank, family bonds, and within- and between-family relations allow her to anticipate what sort of social relationships the infant will have with others.

Just as humans have an innate predisposition to learn language, and nutcrackers an innate predisposition to store and remember the location of seeds, so do baboons and other monkeys have an innate predisposition to recognize other individuals' ranks and social relationships. Males cannot help but keep a running mental tally of changes in other males' ranks, just as Ahla the goat-herding baboon cannot prevent herself from knowing which kid goes with which ewe (Chapter 2). We are still a long way, though, from knowing how malleable this predisposition is, and what mechanisms underlie it. In this respect, those of us who study primate social knowledge find ourselves in a position much like that of psycholinguists in the late 1950s, when Chomsky wrote his critique of Skinner's *Verbal Behavior:* we know that the system we are studying is complex and that its development cannot be explained by simple learning mechanisms alone. But we do not yet understand how it develops in the minds of our subjects. To quote Chomsky's (1959) review, without the specific references to language and substituting *baboon* for *human:*

As far as acquisition … is concerned, it seems clear that reinforcement, casual observation, and natural inquisitiveness (coupled with a strong tendency to imitate) are important factors, as is the remarkable capacity of the baboon to generalize, hypothesize, and "process information" in a variety of very special and apparently highly complex ways which we cannot yet describe or begin to understand, and which may be largely innate, or may develop through some sort of learning or through maturation of the nervous system. The manner in which such factors operate and interact … is completely unknown.

SEVEN

The Social Intelligence Hypothesis

If I be asked by what power the creator has added thought to so many animals of different types, I will confess my profound ignorance.
CHARLES DARWIN, 1838: *NOTEBOOK C*

Does the social intelligence of primates make them smarter than other animals? During the 18th and 19th centuries, both before and after the publication of *The Origin of Species,* scholars debated the merits of different animal minds. At issue was the general question of whether species should be ranked according to their utility to humankind or some other standard (Ritvo 1987). Mental ability seemed a compelling criterion for ranking animals, but even with their admittedly large brains the monkeys and apes did not always come out on top. Some scientists favored the orangutan, whereas others, like George Romanes (1881), argued for the "high intelligence" and "gregarious instincts" of the dog, which gave it a "more complex" psychology than monkeys. In 1883, Arabella Buckley, a friend of A.R. Wallace, published a children's book entitled *The Winners in Life's Race: or, The Great Backboned Family* in which she classified monkeys and apes with insectivores and rodents, rather than at the head of the animal kingdom. The "degenerate gorilla" was described as "equal neither in beauty, strength, discernment, nor any of the nobler qualities, to the faithful dog, the courageous lion, or the half-reasoning elephant" (Buckley 1883; Ritvo 1987).

The social intelligence hypothesis

As we saw in Chapter 2, the notion that baboons' intelligence evolved to solve social problems first appeared in the idiosyncratic writings of Eugene Marais, who alone among his contemporaries observed baboons in their natural habitat. Years later, in 1953, the social intelligence hypothesis reappeared when the British psychologist Michael Chance suggested, on the basis of observations made at the London Zoo, that the near-continuous sexual receptivity of primate females created complex problems in behavioral control and social awareness for males. He postulated that "the enlargement of the neocortex is an anatomical adaptation to [these] circumstances" (Chance and Mead 1953:433). The idea lay dormant for thirteen years. In 1966, the hypothesis was reintroduced by Alison Jolly, one of the first scientists to study primates in their natural habitat. Jolly took a broader perspective than Chance, arguing that "the social use of intelligence" is crucially important to both male and female primates, and that "social integration and intelligence probably evolved together, reinforcing each other in an ever-increasing spiral" (Jolly 1966:504). But Jolly's hypothesis, like Chance's, was largely ignored by a psychological community that believed primate intelligence was best studied by presenting single animals in cages with problem sets comprised of blue squares and red triangles.

The modern renaissance of the social intelligence hypothesis began in 1976, when Nick Humphrey published a short paper entitled "The social function of intellect." Echoing Darwin, Humphrey argued that evolutionary biology and psychology could be combined to reveal the selective factors shaping the evolution of intelligence. He began by noting that natural selection is ruthlessly economical: traits, particularly energetically costly ones, do not evolve unless they serve some function. Years of laboratory testing had shown that monkeys and apes "possess impressive powers of creative reasoning." So what, precisely, were the demands of natural life that had made these animals so clever? Humphrey proposed that "the higher intellectual faculties of primates have evolved as an adaptation to social living" (1976:316) (Fig. 26).

The social intelligence hypothesis does not argue that baboons and other monkeys have little knowledge about their home range, the spatial and temporal distribution of trees, or the behavior of their predators. Indeed, a rigorous comparison of "social" and "nonsocial" skills would be difficult precisely because the two are inextricably entwined. A baboon's ability to acquire the most nutritious food depends, simultaneously, on

Figure 26. A young Sylvia sits next to her mother, Stroppy. Photograph by Joan Silk.

both her ecological knowledge of plants and her skill in competition with others. Her ability to detect and evade predators depends, simultaneously, on both her knowledge of predator behavior and her ability to live cooperatively in a large group, where she benefits from predator detection and defense. Instead, the hypothesis argues simply that all group-living animals confront a multitude of social problems, and that intelligence in primates—and perhaps many other species—must have evolved at least in part because natural selection has favored individuals who are skilled at solving these problems.

Below we review three kinds of evidence that support the social intelligence hypothesis. We then consider several kinds of evidence that might lead us to modify it and reconsider whether primates are unique among animals in their ability to monitor, or eavesdrop upon, other individuals' social relationships.

Social complexity and brain size

Ethological studies have repeatedly shown that natural selection acts on both behavior and the neurological structures that support it. If a nutcracker can improve its survival by hiding pine seeds in the fall and recovering them during the winter, natural selection will simultaneously favor behavior like hiding and searching and an enlarged hippocampus, the area of the brain devoted to spatial memory. Signs of the coevolution of behavior and brain morphology are clearly evident when we compare closely related species. Nutcrackers rely more on stored seeds than do their close relatives, scrub jays (*Aphelocoma californica*), and the nutcracker's hippocampus is correspondingly larger (Kamil et al. 1994).

The coevolution of behavior and brain morphology can also be seen when we compare males and females in certain species. In North American and European songbirds, where males do most of the singing, males also have much larger song control areas in their brains than do females (reviewed in Brenowitz and Kroodsma 1996). But in many tropical songbirds, where males and females sing equally, the sizes of male and female song control areas are more alike (Brenowitz and Arnold 1986). Song control areas in the brain have thus coevolved with behavior. Male baboons, who can increase their reproductive success by winning aggressive contests with rivals, have evolved large body size, large canines, and competitive vocal displays. Female baboons, whose reproductive success is much less influenced by fighting, have evolved neither the same morphology nor the same behavior.

Applying this logic to the social intelligence hypothesis, we might predict that species living in large, complex societies should exhibit both more sophisticated social knowledge and bigger brains than solitary species or species that live in small, monogamous groups. Alternatively, if large brains have evolved for some other reason, there should be no predictable relation between brain size, group size, and social intelligence. So why have large brains evolved? What are big brains for?

Across the animal kingdom, brain size increases with body size. Despite this common scaling principle, however, brain size–to–body weight ratios differ from one taxonomic group to another. Among mammals, primates have brains that are larger, on average, than the brains of similar-sized, nonprimate mammals.

Within the primate order, the picture is more complex. Martin (1990) uses the index of cranial capacity (ICC) as a means of comparison. The ICC is the ratio of a species' actual brain volume to that which would be expected for an animal of the same size if it were a basal insectivore—the basal insectivores being taken as a standard point of comparison. The ICC thus measures the extent to which a species diverges from the "typical" pattern for this group (see also Fuster 1997; Rilling and Insel 1999). Baboon brains average 177 cc in volume, giving them an ICC of 7.3, whereas chimpanzee brains average 393 cc, with an ICC of 8.2. Although this might suggest that all ape brains are relatively larger than all monkey brains, there is considerable variation. The capuchin monkey (*Cebus* sp.), a New World primate, has an ICC of 11.7, whereas the gorilla, a great ape, has an ICC of 5.5. Overall, great apes closely fit the same scaling relationship as monkeys for brain:body size relations: their brain sizes are approximately what would be expected for a monkey scaled up to the appropriate body size (Martin 1990; Rilling 2006).

The brains of great apes may, however, differ from those of other primates on some qualitative, structural dimensions. Ape brains appear to exhibit a comparatively greater elaboration of the cerebellum and frontal lobes (Rilling 2006). In chimpanzees, a larger proportion of the brain is devoted to neocortex than in monkeys, which in turn have proportionately more neocortex than prosimians (Passingham 1982; Rilling and Insel 1999). Within the neocortex, ape (and especially human) brains have a particularly enlarged prefrontal cortex, an area known to be involved in many forms of abstract thought, decision making, rule learning, and reasoning about others' mental states (Deacon 1997; Fuster 1997; Rilling 2006).

Increases in the size of primate brains have come despite the fact that brain tissue is metabolically very costly. As we mentioned in Chapter 1, by one estimate the human brain uses energy at the same rate as the leg muscles of a runner during a marathon. Furthermore, large brains take a long time to develop. Monkeys experience a longer period of juvenile dependence and live longer than most other mammals of similar body size. Apes, in turn, experience a longer period of juvenile dependence and live longer than monkeys. Both of these life history traits are correlated with large brain size (van Schaik 2004). However, a prolonged period of prereproductive life is risky—you could easily die before you are able to reproduce. Large brains must therefore provide individuals with benefits that more than compensate for these costs. What are these benefits? When the question is applied to human evolution, answers typically focus on the adaptive advantages of technology (initially, stone tools) and language. But most monkeys rarely, if ever, use tools and lack language entirely, yet their brains are significantly larger than those of similar-sized mammals. At least in the case of monkeys, some other selective pressure must be at work.

Early studies found that brain weight:body weight ratios were higher in primate species with larger home ranges and larger in species that were fruit-eating or omnivorous than in leaf-eating species (Clutton-Brock and Harvey 1980; Milton 1988). These relationships suggested that fruit-eating primates face special problems in learning and memory because they depend on widely dispersed food that is ephemeral in both space and time. Other data, however, argue against this explanation. In an analysis that involved many primate species, Barton and Dunbar (1997) showed that group size, not home range size, best predicted the size of a species' neocortex. They argued that group size is a good proxy for social complexity and concluded that primate brains have evolved in response to the demands of social life. Group size is important in part because, as we discussed in the previous chapter, the calculation of other individuals' social relationships becomes increasingly daunting as group size increases. Social competition might therefore offer one explanation for primates' unusually large brains. The hypothesis seems particularly compelling in the case of monkeys (if not apes), because monkeys rarely use tools but often live in relatively large groups.

So why do monkeys live in large groups? Some hypotheses stress the benefits of group life for predator detection and defense (e.g., van Schaik 1983), whereas others stress the need to defend food resources (e.g., Wrangham 1980). Still others suggest that both factors are important (e.g., Sterck et al. 1997). Whatever the reasons, these socioecological hypotheses propose that selection favoring large groups has resulted in societies that contain many different families, with social relationships simultaneously competitive and cooperative. This, in turn, places strong selective pressure on individuals' skills in managing social relationships.

Other hypotheses place less emphasis on group size and more emphasis on the details of primate feeding ecology. Kaplan and colleagues (2000) suggest that the need to forage in a complex three-dimensional environment for ephemeral fruit may have favored the evolution of cognitive abilities that served as preadaptations for the evolution of complex social relationships. Fruit-eating primates have relatively enlarged areas of the brain devoted to visual acuity, binocular vision, and color vision (Barton 1998). Thus, the cognitive demands of a frugivorous diet might have favored the evolution of large brains that could, in turn, support complex societies. On the other hand, it could also be argued that the need to process rapidly changing social interactions, such as those that occur during alliances, demands rapid visual processing and integration of information about individuals' behavior and gaze direction. Thus, social interactions might place just as strong, if not stronger, demands on vision and cognition.

It may never prove possible to determine whether social pressures, ecological pressures, or both drove the evolution of large brains in primates. Indeed, other selective pressures, such as social learning and technological innovations, may have exerted far greater influences, particularly in the ape and hominid lines (Reader and Laland 2002). Brain size is correlated with group size, but the causal origins of this correlation remain elusive. It even remains unclear whether primates differ from other animals in their "social intelligence."

Brain specializations for social stimuli

Preliminary evidence in support of the social intelligence hypothesis as it applies to primates comes from the existence of brain areas specialized to deal with social stimuli. Monkeys, for example, have "face cells" in the temporal cortex that respond at least twice as vigorously to faces or components of faces (like eyes or mouths) than to other complex visual stimuli (Tsao et al. 2003, 2006). Face cells are surprisingly specialized. Those in the inferior temporal cortex (IT) seem most important for processing facial identity, whereas those in the superior temporal sulcus (STS) seem most important for processing facial expressions. IT and STS are extensively interconnected and probably share face-specific information (Weiss et al. 2002; Ghazanfar and Santos 2004). Face cells in STS respond not only to facial expressions but also to the direction of an individual's head orientation and gaze. Their response is greatest when head orientation and gaze direction are congruent, less strong when they are incongruent (Emery and Perrett 2000; Jellema et al. 2000; Perrett et al. 1992; Eifuku et al. 2004). The STS of rhesus macaques also includes neurons that fire when the monkey observes an individual walking, turning his head, bending, or extending his arm (Perrett et al. 1990). Particularly intriguing are "mirror neurons" in the inferior parietal lobule that show elevated activity both when the subject monkey executes a specific grasping action and when the monkey observes a human or other monkey execute a more or less similar grasp (Rizzolatti and Craighero 2004).

Finally, monkeys—like humans—process their own species' vocalizations in ways that are measurably different from the way they process other auditory stimuli. As is well known, most humans exhibit lateralization in their perception of speech. Because language is typically processed in the left hemisphere of the brain and the left hemisphere has its primary connections to the right side of the body, most humans are better at making rapid assessments of words and sentences when they hear them through their right ear. Their right-ear advantage disappears, however, when they are asked to identify other auditory stimuli (see Rogers and Andrews 2002 for review). Like humans, monkeys display a left-brain, right-ear advantage when processing their own species' vocalizations, but not when processing other sounds (Petersen et al. 1978; Hefner and Hefner 1984; Weiss et al. 2002; Poremba et al. 2004).

Specialized cells and unique brain mechanisms for dealing with social stimuli do not, by themselves, confirm the social intelligence hypothesis. Regardless of their degree of social complexity, all animals' brains respond selectively to stimuli produced by members of their own species. Solitary frogs respond selectively to their own species' calls, and monogamous birds respond selectively to their own species' song. But the special responsiveness of the monkey's brain to monkey voices, faces, gaze direction, and actions are just what we would expect if natural selection had acted with particular force to favor individuals skilled in solving social problems. Particularly intriguing is evidence that the STS and mirror neurons are also highly sensitive to intentional, goal-directed behavior, indicating that they may help monkeys assess other individuals' intentions. We discuss this issue further in Chapter 8.

Social versus ecological "intelligence"

For 11 years during the 1970s and 1980s, we studied the social behavior of vervet monkeys in Kenya's Amboseli National Park (Fig. 27). Vervets

Figure 27. A matriline of vervet monkeys. Photograph by Dorothy Cheney.

live in smaller groups than baboons (10–25 on average), but their social organization is similar. Like baboons, they also display impressive knowledge of other individuals' dominance ranks and kinship relationships. They practice strategic alliances formation, curry favor with the members of high-ranking families, and keep track of who has been nice to them in the recent past (Cheney and Seyfarth 1990).

But we were equally struck by what the vervets seemed *not* to know. Take animal tracks, for example. Vervets in Amboseli regularly fall prey to pythons (*Python sebae*), large constrictors that hunt by concealing themselves in long grass or thick bush. Occasionally, though, the python's track gives away its location. When crossing open ground, pythons lay distinct, wide, straight tracks that cannot be mistaken for those of any other snake. Local humans recognize them easily, and it is relatively easy to find a concealed python by following its track.

Yet vervets seem unable to recognize that a fresh python track signals danger. On more than one occasion we watched as a vervet calmly followed a fresh track into a bush, only to leap out in surprise and alarm when it encountered the actual snake. The vervets' ignorance of python behavior was striking, because their daily life had certainly provided them with ample opportunity to watch pythons move across open areas laying down a track, and to associate this distinctive track with the animal itself. The association between a fresh track and a python was as statistically reliable as the association between two vervets in the same family who groom each other at high rates, but evidently more difficult for the vervets to learn.

Other gaps in the vervets' performance revealed that their knowledge outside the social domain was not what one would expect. Vervets are frequently attacked by leopards, which are abundant in Amboseli. Unlike other African cats, leopards carry their prey into trees, where it cannot be stolen by lions and hyenas. We had often seen vervets giving alarm calls as they watched a leopard haul the carcass of a gazelle into a tree. They clearly had the opportunity to learn that a carcass in a tree signaled the likely proximity of a leopard. But when we placed a stuffed carcass of a Thompson's gazelle (*Gazella thomsonii*) in a tree near the vervets' sleeping site before dawn, the monkeys showed no response. We might just as well have put a suitcase in the tree. It was not that the carcass was unrealistic; it did fool a tour bus operator. What was striking was that the monkeys failed even to show any curiosity about it (Cheney and Seyfarth 1985a). (We should note here that the local baboons also ignored the carcass. In fact, baboons are not even especially alarmed when they see the carcass of a *baboon* cached in a tree by a leopard. Although they look at it with curiosity, they seem not to recognize that the carcass is associated with a leopard.)

This is not to say that vervet monkeys are oblivious to the signals provided by other species. They respond strongly to the alarm calls given by ungulates and birds, and even differentiate among the alarm calls given by birds to avian and terrestrial predators (Cheney and Seyfarth 1990). Similarly, they differentiate between the lowing of wildebeest (*Connochaetes taurinus*), which pose no threat, and the lowing of cows, which signal the approach of Maasai herders. Vervets are, therefore, not completely ignorant about their physical world. With training, they could also certainly learn to recognize the association between a python and its track, or a leopard and a carcass in a tree. But in the absence of such training they seem predisposed to attend to events in the social world in a way that they do not attend to events in the physical world. Vervets have a kind of laser-beam acuity; they make good psychologists but poor naturalists.

Such "attentive biases" are common among animals. Chickadees, for example, store seeds in the fall and, like nutcrackers, have excellent spatial memories. In one experiment, captive chickadees were trained to find food in a particular box located in an array of boxes positioned on a wall. The food might be located in a yellow box that was positioned third in line after a red box and a blue box, but before a green box. Once they had learned to choose the yellow box, the array was shifted along the wall and the boxes were scrambled. Now, say, the yellow box was first in line. When released into the aviary, most chickadees ignored both the color of the box and its relative position in the array. Instead, they flew to the box that was located at the same place on the wall where the yellow box had originally been (Brodbeck 1994). When learning which box contained food, the chickadees had apparently paid most attention to the box's location on the wall, ignoring both its color and its relative position in the array. In contrast, juncos (*Junco hyemalis*), close relatives of chickadees that do not store food, chose the box according to its color. Chickadees are not completely oblivious to cues like color. When trained, they can learn to choose boxes according to their color. Instead, they seem biased to pay attention first and foremost to spatial cues. Other food-storing birds like nutcrackers display a similar bias (Olson et al. 1995). In a comparable way, vervets seem biased to be particularly attentive to social events.

Critique of the argument thus far

To this point, the argument in favor of the social intelligence hypothesis, at least as it applies to primates, can be summarized as follows. Compared with other mammals, primates have larger brains relative to their body weight. Within the primate order, brain size is strongly correlated with group size. Since the complexity of an individual's social relationships increases exponentially with increasing group size, large brains seem to have evolved in response to the demands of social life. Consistent with this hypothesis, we find that the primate brain contains many areas specialized for dealing with social stimuli, like the faces, movement, and voices of members of their own species; that baboons and other monkeys recognize each other's dominance rank and social relationships; that female baboons' social relationships are correlated with reduced stress and increased reproductive success; and that vervet monkeys' knowledge of their social companions is impressive, whereas their knowledge of some ecological relations is underwhelming.

The argument can be challenged on at least four grounds. First, we have been discussing "intelligence" in primates and other animals without any attention to the behaviorists' critiques mentioned in the last chapter. Knowledge of other individuals' relationships might be acquired through relatively simple associative processes—processes that could easily be duplicated by even small-brained species.

Second, because primates have relatively larger brains than other species and large brains are presumed to have evolved to deal with social complexity, it follows that primate societies must in some fundamental way be more complex than those of other species. But we have thus far presented no evidence to support this view. Dolphins (*Tursiops truncates*), elephants (*Loxodonta africana*), spotted hyenas (*Crocuta crocuta*), and pinyon jays (*Gymnorhinus cyanocephalus*)—to name just a few species—also live in complex societies comprised of multiple families and stable dominance hierarchies. Do they too recognize and monitor other individuals' social relationships?

Third, the argument assumes that skills in recognizing social relationships have evolved in response to the challenges of living in large groups. Yet, as we will describe, recent studies have revealed similar skills in social intelligence in solitary animals and animals living in small family groups, such as monogamous birds. How do we reconcile these data with the social intelligence hypothesis?

Finally, within the primate order, social learning, innovation, and tool use are strongly correlated with brain size but not with group size.

In particular, chimpanzees and orangutans have larger brains than monkeys and use and manufacture tools more routinely than monkeys, but live in relatively small groups. Indeed, orangutans are frequently solitary. These relationships suggest that many of the selective pressures favoring enlarged brains in apes and humans may have been technological rather than social. We reevaluate the "social intelligence" hypothesis in light of this evidence at the end of the chapter.

Are primates different?

Primates have, on average, larger brains for their body size than other vertebrates. Dunbar (2000, 2003) argues that this came about because primate social groups are not only larger but also more complex than those in other taxa. Primate groups are typically composed of many reproductively active males and females, and individuals interact with both kin and non-kin in both competitive and cooperative contexts. Such social complexity may place strong selective pressure on the ability to recognize other individuals' ranks and social relationships.

Some comparative tests of captive apes, monkeys, pigeons, and other animals suggest that primates are more adept than nonprimates at classifying items according to their relative relations. In oddity tests, for example, a subject is presented with three objects, two of which are the same and one of which is different, and asked to choose the object that is different. Monkeys and apes achieve high levels of accuracy, even when tested with novel stimuli (reviewed by Tomasello and Call 1997; Shettleworth 1998). In all cases, subjects' performance suggests the use of an abstract hypothesis, because concepts like *odd* specify a relation between objects independent of their physical features. In a similar manner, the concept *closely bonded* can be applied to any two individuals and need not be restricted to individuals that look alike. Although many animals, including honeybees (*Apis mellifera;* Giurfa et al. 2001), can learn to solve "delayed match-to-sample" (pick the stimulus that is the same as the previous one) and "delayed non-match-to-sample" (pick the stimulus that is different from the previous one) tasks, primates typically learn faster and generalize more accurately to novel stimuli (e.g., Harlow 1949; Strong and Hedges 1966; Wright et al. 1984). Furthermore, primates can be taught rules about how to respond: for instance, "If the tray is green, pick the odd item; if it's red, pick the same one."

Baboons and chimpanzees can also learn to make abstract judgments that involve comparing one relation with another (Premack 1983; Oden et al. 1988; Fagot et al. 2001). In one study, the language-trained chimpanzee Sarah was tested on her ability to reason analogically. When Sarah was shown a lock and a key and asked to pick the appropriate object to accompany a can and complete the same relation, she correctly chose a can opener. She therefore completed the analogy "key is to lock as can-opener is to can." (This test will doubtless bring back dark memories to all American readers who remember the analogical reasoning portion of the SAT featuring questions like "flounder is to telephone booth as yak is to (a) democracy, (b) the Vietnamese pot-bellied pig, (c) summer, (d) Kant's *Critique of Pure Reason*, (e) none of the above, (f) all of the above.")

The ability to make judgments based on relations among items has been demonstrated more often in primates than in other animals, and primates seem to recognize abstract relations more readily than at least some other animals. Although it is possible to train pigeons (*Columba livia*) to recognize relations such as *same*, the procedural details of the test appear more critical for pigeons than they are for monkeys. Rather than attending to the relations among stimuli, pigeons seem predisposed to focus on the physical features of the stimuli and to form item-specific associations (reviewed by Shettleworth 1998). Similarly, in tests of transitive inference, monkeys and apes appear to acquire a representation of serial order that allows them to rank items even when some items in the list are missing (D'Amato and Colombo 1989; Treichler and van Tilburg 1996). So, for example, having learned the series A > B > C > D, monkeys have little difficulty recognizing that B > D. In contrast, many—but not all—birds seem to attend primarily to the association between adjacent pairs, which limits their ability to add or delete items from a list (von Fersen et al. 1991; Zentall et al. 1996).

Recent experiments by Earl Miller and colleagues have begun to elucidate the neural basis of abstract judgments in rhesus macaques. In a typical test, subjects first saw a picture and received a cue: as they viewed the picture they either received a drop of juice or heard a tone. Then they were shown a second picture. If they had been given juice, they were to respond only if the second picture was the same as the first. If they had heard a tone, they were to respond only if the second picture was different from the first. Thus the monkeys had to learn the abstract rule "juice drop means *same* and tone means *different*" and apply this rule regardless of what the pictures actually showed. The monkeys readily learned to do this. Underlying their behavior was neural activity in the prefrontal cortex (PFC), where individual nerve cells appeared to exhibit rule specificity. Some cells sharply increased their firing when the *same* rule was in force, whereas others increased their firing when the *different* rule was in force. The cells' selective firing could not be explained by the physical features of the pictures because these were different in different trials. Nor could selective firing be linked to the monkeys' anticipation of their response, because at the time they received the cue and learned which rule was in force the monkeys did not yet know how they would respond. Instead, the selective neurons seemed to function in the encoding of an abstract rule (Freedman et al. 2001; Wallis et al. 2001; Miller et al. 2002). Complementing these data, recall that the relatively greater size of primate brains is particularly pronounced in the prefrontal cortex.

Taken together, these data argue that the primate brain—particularly the prefrontal cortex—and primates' cognitive skills—particularly the ability to make abstract judgments—have evolved together, presumably in response to the demands of a socially complex society. As a result, modern nonhuman primates have both larger brains and greater cognitive abilities than other animals. We might be tempted to conclude that cognitive skills in primates are qualitatively and hence fundamentally different from those in all other animals.

Social cognition in gregarious mammals and birds

This conclusion, however, may be premature. If the ability to recognize and monitor other individuals' social relationships confers a selective advantage, we should expect to find evidence of social intelligence and increased brain size not just in primates but also in any animal species that lives in large social groups, particularly those that contain individuals of different dominance ranks and varying degrees of genetic relatedness. Conversely, selection should have acted less strongly on social intelligence and brain size in solitary species and species living in small, family groups. If true, the ability to recognize the close associates of others should be evident in nonprimate species like hyenas and dolphins and lacking or less highly developed in the less social apes, including gorillas and orangutans. Studies to test this hypothesis are only beginning to be conducted. Indeed, one of the great lacunae in cognitive studies of apes (including chimpanzees) is the absence of any research on apes' ability to monitor other individuals' social relationships.

There is, in fact, good evidence that social complexity and large brains have coevolved in nonprimate species as well as in monkeys and apes. As we mentioned earlier, primate species that live in large groups have a relatively larger neocortex than those that are solitary or live in small groups. A similar relation is found in carnivores (Barton and Dunbar 1997), toothed whales (Connor et al.1998; Marino 1998), and ungulates (Perez-Barberia and Gordon 2005), supporting the hypothesis that sociality has driven the evolution of large brains in many taxonomic groups. Differences in social complexity may exert their effect even in species that lack a cortex entirely. In paper wasps (*Polistes dominulus*), for example, there is a significant increase in the size of the antennal lobes and collar (a substructure of the calyx of the mushroom body in the insect brain) in females that nest colonially, with other queens, as opposed to solitary breeders (Ehmer et al. 2001). This increase in neural volume may have been favored because sociality places increased demand on the need to discriminate between familiar and unfamiliar individuals and to monitor other females' dominance and breeding status. Changes in brain size occur even within individuals, according to the size of their behavioral repertoires. The brains of queen ants are significantly smaller than those of virgin females during their nuptial flight. Queen ants are also much less socially active and much less reliant on vision (Julian and Gronenberg 2002). Clearly, therefore, neural correlates of sociality are not restricted to higher mammals.

Given their relatively large brains, it is not surprising that highly social nonprimate mammals also display sophisticated knowledge of other individuals' social relationships. When competing over access to females, male dolphins form dyadic and triadic alliances with specific other males, and allies with the greatest degree of partner fidelity are most successful in acquiring access to females (Connor et al. 1992, 1999, 2001). The greater success of high-fidelity alliances raises the possibility that males in newly formed alliances, or in alliances that have been less stable in the past, recognize the strong bonds that exist among others and are more likely to retreat when they encounter rivals with a long history of cooperative interaction.

Similarly, spotted hyenas live in baboon-like social groups made up of matrilines in which daughters inherit their mothers' dominance ranks (Smale et al. 1993; Engh et al. 2000). Holekamp and colleagues (1999) played recordings of cubs' *whoop* calls to mothers and other clan members. Like vervets and baboons, hyena females responded more strongly to the calls of their offspring and close relatives than to the calls of unrelated cubs. In contrast to vervets and baboons, however, unrelated animals did not look at the cubs' mothers.

One explanation for these negative results is that hyenas are unable to recognize other individuals' kin relations, despite living in social groups that are superficially similar to those of many primates. It is also possible that hyenas are simply uninterested in the calls of unrelated cubs. In fact, hyenas' patterns of redirected aggression suggest that they do recognize other individuals' kin (or close associates). Like monkeys, hyenas sometimes "redirect" aggression toward other, previously uninvolved animals after they have been in a fight. When redirected aggression occurs, hyenas are most likely to attack a relative of their former opponent (Engh et al. 2005).

Hyenas also seem to make transitive inferences about other individuals' dominance ranks. When competing over meat, hyenas often solicit support from other individuals, and they typically solicit aid from allies that are dominant to their opponent (Fig. 28). Similarly, when choosing to join an ongoing skirmish, a hyena that is dominant to both of the contestants almost always supports the higher-ranking of the two (Engh et al. 2005). If the hyena is intermediate in rank between the two opponents, it inevitably supports the dominant. These data provide the first evidence that individuals in a nonprimate species base their decision to join an alliance on both their own and the other individuals' rank relations. Like monkeys, hyenas appear to monitor other individuals' interactions and learn about other animals' ranks from their observations.

Figure 28. Two hyenas form a coalition against a third. Photograph by Kay Holekamp.

Ultimately, the best way to test whether sociality has favored specialized cognitive skills is to conduct comparative tests on closely related social and nonsocial species, similar to the ones that have been conducted on bird species that do and do not store food. For example, Clark's nutcrackers have prodigious spatial memory for storing and recovering food and a bias to attend to spatial, rather than nonspatial, cues. But although they outperform closely related jay species in radial maze and cache retrieval tasks, they are less accurate than other jays when the tests require memory of color (Olson et al. 1995). They are also relatively inattentive to social cues.

Clark's nutcrackers are relatively asocial. Although they form monogamous groups during the breeding season, during the fall and winter they search for and store seeds alone. By contrast, their close relatives, Mexican jays (*Aphelocoma ultramarina*), live in highly structured flocks numbering several dozens of birds. In addition to storing their own seeds, Mexican jays also pilfer from the caches of others. In experiments where birds could recover seeds from sites that either they themselves had created or they had observed another bird create, nutcrackers were more accurate at recovering their own caches than at recovering other birds' caches. Mexican jays, on the other hand, were as accurate at recovering seeds from caches they had observed another bird create as ones that they had created themselves (Bednekoff and Balda 1996). Perhaps because they almost always store seeds alone, nutcrackers are not very attentive observers, and they more accurately remember what they do than what others do. For the more social Mexican jays, however, it pays to spy on your neighbor. As a result, Mexican jays attend to and remember not just what they do but also what others do (see also Lefebvre et al. 1996 for similar data on gregarious pigeons and territorial doves). We return to this interesting question in Chapters 8 and 9.

The social intelligence hypothesis predicts that species living in large social groups should be able to track other individuals' relationships and ranks more accurately than closely related species living in small family groups. Alan Bond, Al Kamil, and their colleagues tested this hypothesis on two species of jay with markedly different social organizations.

Pinyon jays are highly gregarious—they are sometimes referred to as "avian baboons." They live in stable flocks of 50 to 500 birds, each containing individuals that are linked by kinship and arranged in a linear dominance hierarchy. By contrast, western scrub jays (like nutcrackers) live either alone or in small monogamous family groups. The two species are closely related and sympatric throughout much of Colorado and Arizona. The birds were tested on their ability to make transitive inferences—that is, to recognize that if A is dominant to B and B is dominant to C, then A must be dominant to C. Transitive inference is crucial to the recognition of relative rank in a dominance hierarchy.

In the first experiment, each bird was presented with two stimuli, A and B (denoted by different colored circles), and rewarded for selecting A. Once the bird had reached a certain level of performance, it was presented with a novel pairing, B and C, and rewarded for selecting B (note that B is the incorrect answer when paired with A, but the correct answer when paired with C). After reaching criterion on the B/C pairing, they were tested with C/D (C was rewarded), D/E (D rewarded), and E/F (E rewarded). Pinyon jays performed significantly better than scrub jays, especially as more items were added to the list.

In the next experiment, the birds were tested with nonadjacent stimuli like B and D. Both species performed well, especially on nonadjacent pairs like B/D that were drawn from the top part of the hierarchy. But when the birds were tested with nonadjacent stimuli, like D/F, toward the bottom of the hierarchy, a striking difference emerged: scrub jays responded quickly but were often wrong, whereas pinyon jays took much longer to respond but were usually correct. Apparently, during training the scrub jays had memorized each combination of pairs, whereas the avian baboons, the pinyon jays, had memorized the ranked list. As a result, although the pinyon jays seemed to have to "recite" the entire list in their heads before making a choice, they were more accurate than scrub jays at recognizing the relative relation between nonadjacent pairs (Bond et al. 2003).

A subsequent experiment has shown that pinyon jays use transitive inference to calculate their own dominance status relative to that of a stranger they have observed interacting with their group-mates. In this study, four jays were placed in each of two cages and allowed to form their own dominance hierarchies. We will call the dominance hierarchy in the first group A > B > C > D and the one in the second 1 > 2 > 3 > 4. The cages were kept in separate rooms, so the jays were unfamiliar with the members of the other group. A bird (say, bird 3) was then temporarily removed from its group and allowed to witness two paired encounters (Fig. 29). In the first A dominated B; in the second B dominated 2 (note that only bird 2 was familiar to bird 3). Next, bird 3 was placed in the same cage as bird B. In all cases, bird 3 deferred to B. The bird acted as if he had made the transitive calculation "B may be subordinate to A, but he's clearly dominant to 2. Given that 2 is dominant to me, I'd better be submissive to B." The bird could only have made this

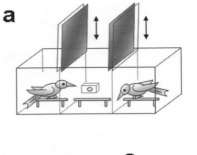

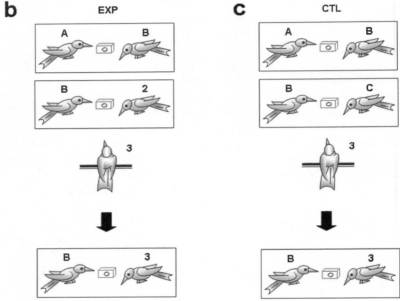

Figure 29. The protocol for tests of transitive inference in pinyon jays. (a) All dyads in two social groups were first tested in competitive interactions to establish their relative dominance ranks. In one group, A > B > C, etc. In the other group, 1 > 2 > 3, etc. Next, a bird (e.g., bird 3) from one group observed a staged competitive encounter between two other contestants. If bird 3 was in the test condition (b), he observed A > B and B > 2, where only 2 was familiar to 3. If bird 3 was in the control condition (c), he observed A > B and B > C, where all three birds were unfamiliar to 3. In either case, bird 3 was then tested with bird B. Bird 3 was significantly more likely to behave subordinately to B in the test condition. Figure provided by G. Paz-y-Miño.

calculation by taking a non-egocentric perspective and recognizing the relative ranks that exist among others (Paz-y-Miño et al. 2004).

In less social settings, many gregarious species of birds and mammals can make similar relational distinctions. For example, the African gray parrot (*Psittacus erithacus*), Alex, is reported to make explicit same/ different judgments about sets of objects (Pepperberg 1992). Sea lions and dolphins have been taught to respond to terms such as *left, right, dark,* and *bright* that require them to assess relations among a variety of different objects (Schusterman and Krieger 1986; Schusterman et al. 1993; Herman et al. 1993a; Mercado et al. 2000). And a number of species, including parrots and rats, are able to assess quantities, suggesting that relatively abstract concepts of number and transitivity may be pervasive among animals (Pepperberg 1994; Church and Meck 1984; Capaldi 1993; Suzuki and Kobayashi 2000; reviewed by Shettleworth 1998).

Finally, birds and nonprimate mammals like dogs seem to be at least as adept as monkeys and apes at gleaning information from other individuals' direction of gaze, at recognizing other individuals' perspectives and motives, and at concealing their behavior from others. The ability to read other individuals' minds—if only at a very rudimentary level— constitutes a crucial component of social intelligence. We discuss this question at length in Chapters 8 and 9.

Taken together, the data from studies of dolphins, hyenas, and several species of birds raise the possibility that monkeys appear to have a greater capacity to recognize other individuals' relationships only because they have received more attention than nonprimate species living in similarly large groups. Once this imbalance in research has been redressed, the difference between primates and other animals will disappear, to be replaced by a difference that depends primarily on group size and composition.

Social cognition in more solitary species

It is also possible that neither phylogeny nor group size has determined an individual's ability to gain information about others. There may, in effect, be no substantive differences across species in "social" intelligence. After all, much information about other individuals' social relationships can be acquired through relatively simple processes of association and transitivity. If animal A outranks B, and B outranks C, it is not too difficult to conclude that A will outrank C. Highly social species like baboons, hyenas, and pinyon jays might appear to excel in their ability to recognize other individuals' relative ranks only because their large social groups allow them to display this knowledge. Solitary animals, or animals that live in small family groups, might be just as skilled but fail to display their knowledge because the opportunity to monitor interactions among many other individuals rarely arises.

Recent research on "eavesdropping" by birds and fish has shown that animals living in small social groups are indeed capable of acquiring detailed information about other individuals' relative dominance or attractiveness as a mate. Like their North American relatives the chickadees, great tits (*Parus major*) in Denmark form monogamous pairs in which the male defends his territory against other males by singing. When an intruder encroaches onto a male's territory, the resident typically engages in a singing contest with his rival. In great tits, dominance takes the form of interruption. If a male—either the territory holder or his challenger—feels that he can dominate his rival, he will challenge him by singing over his song. Subordinate males, in contrast, will politely remain silent when their rival is singing.

To test whether male tits acquire information about potential opponents by eavesdropping on other males' interactions, Peake and colleagues (2002) used playback experiments to simulate an encounter between a male subject and an intruder (A). A loudspeaker was placed onto the subject's territory and A's song was played. (Note that A is not a real bird, only a vocal representation of one.) When the subject answered with his own song, a computer manipulated the timing of A's response in such a way that in some contests A won the singing match, while in others the subject did. Fifteen minutes after this playback, the experimenters simulated a singing contest between A and another strange bird, B, outside the subject's territory by broadcasting A's and B's songs from two loudspeakers. In some of these simulated contests, A won the contest; in others B won. The loudspeaker was then moved back onto the subject's territory and B's song was broadcast to the subject.

If the subject had previously dominated A, and A had dominated B, the subject responded only weakly to B's intrusion. He appeared to regard B as someone who posed little threat. But if A had dominated the subject and B had dominated A, the subject responded to B's intrusion by singing at a high rate, as if he recognized that B represented a significant challenge to his territory. Subjects behaved as if they had acquired information about B's dominance relative to their own by noting B's relationship with A and integrating this information with their own experience with A (Peake et al. 2002). They behaved, in other words, as if they were guided by the transitive rule "if I can dominate A, and A can dominate B, then B is no threat to me."

Eavesdropping on the singing contests of other males allows males to assess a rival's status without engaging in a potentially costly conflict.

It also allows females to assess their mate's relative dominance. In an experiment conducted on Canadian chickadees that parallels the study on great tits, females were given the opportunity to eavesdrop on simulated singing contests between their mate and a neighboring male. In some playbacks, the mate appeared to win the contest; in others, he appeared to lose. Subsequent paternity tests on the pairs' offspring showed clearly that females were paying attention. The males who had won the simulated encounters fathered all the chicks in their nest. In contrast, the nests of males who had lost their encounters contained many chicks that had been fathered by the neighboring male (Mennill et al. 2002).

Fish, too, eavesdrop on other individuals' competitive displays when assessing their chances of beating an opponent. In one experiment, male Siamese fighting fish (*Betta splendens*) observed an aggressive interaction between two unfamiliar males from behind a transparent partition. Observers subsequently avoided the winner of the contest at a significantly higher rate than the loser, suggesting that they had gained information about the two fishes' fighting ability through their observations (Oliveira et al. 1998).

Even relatively unsocial species, then, have a sophisticated knowledge of other animals' relations. This behavior is clearly adaptive, because it allows listeners to assess a rival's skills at very little cost to themselves. As we mentioned earlier, however, it is difficult to ascertain whether the social intelligence of solitary or monogamous species is equivalent to that of more social species, because under natural conditions these less social species simply do not have the opportunity to monitor the ranks of more than two or three other individuals. Comparative tests conducted in the laboratory, like those of Bond et al. (2003) on pinyon and scrub jays, lend some support to the hypothesis that social complexity is correlated with superior performance in some cognitive tasks. But more data are needed from both natural and laboratory studies before we can make any definitive conclusions about cognitive differences between primates and other animals, or between species living in large as opposed to small groups.

Reconciling "social" and "technological" intelligence

Finally, we must consider the possibility that large brains in apes and humans—if not in monkeys—evolved as a result of selective pressures favoring social learning and technological innovation rather than skill in social interactions. Recently, Reader and Laland (2002) accumulated data on 116 primate species and looked for evidence of innovation (defined as apparently novel solutions to environmental or social problems), social learning (the acquisition of information from others), and tool use. They assumed that the frequency of such behaviors, corrected for the amount of time that had been devoted to studying each species, would provide a useful measure of a species' behavioral flexibility and that behavioral flexibility is a good measure of intelligence. They found significant positive correlations between brain size and all three behaviors. There was also a close relation between innovation and social learning. Group size, however, was not closely related to any of these behaviors. The lack of a strong correlation between group size and innovation was due primarily to three primates: New World capuchin monkeys, chimpanzees, and orangutans (Fig. 30). All three have large brains, use tools in a variety of contexts, but live in comparatively small groups—in the case of orangutans, sometimes no group at all. These correlations suggest that tool use and behavioral flexibility, not the complexity of social groups, have driven brain evolution in apes and humans (Reader 2003; for a similar argument, see Kaplan et al. 2000; van Schaik 2004).

Figure 30. As her infant looks on, a chimpanzee in the Tai Forest of the Ivory Coast uses a rock to crack open a palm nut. Photograph by Roman Wittig and Cathy Crockford.

Reader and Laland's results remind us that ecological and social skills are difficult to distinguish in present-day species and unlikely to have played entirely separate roles during evolution. Social learning, after all, can help individuals to acquire food, and tool use can have social as well as ecological benefits. Jane Goodall (1968), for example, describes a male chimpanzee who rose in rank after he learned to bang together garbage can lids in dominance displays. Because innovation and tool use are highly dependent on social learning, perhaps the most appropriate conclusion to make is that "social" and "technological" intelligence should not be contrasted as alternatives, but as selective forces that are inextricably linked. In any case, the absence of any data on apes' knowledge of other individuals' social relationships makes it impossible to contrast the two sorts of intelligences in apes.

Another hypothesis—first proposed by Alison Jolly (1966) and one to which we return to at the end of this book—argues that the technological and innovative skills that we see in rudimentary form among chimpanzees (and hyperbolically so in humans) have their roots in the selective forces that originally favored the evolution of social skills. Technological innovations require planning and the kind of "what-if" contemplations that can arise only through introspection. The propagation of innovative technology demands, in turn, the ability to recognize other individuals' goals and motives, to imitate, and in at least some cases also to teach. Chimpanzees routinely manufacture and use simple tools. They also show evidence of planning and imitation. Baboons and other monkeys rarely if ever manufacture tools, but they may have a limited capacity to access and monitor their own knowledge and to attribute mental states to others (Chapters 8 and 9). Indeed, inferences about other individuals' intentions—or at least their intention to behave in certain ways—guide almost every component of baboons' social behavior, including especially their vocal communication. In addition to placing computational demands on their participants, therefore, social groups may require individuals to make inferences about other group members' intentions and to plan alternative scenarios for future behavior. It seems highly plausible, as Reader and Laland suggest, that the "ability to learn from others, invent new behaviors, and use tools may have played pivotal roles in primate brain evolution" (2002:4436). It also seems likely, though, that these skills piggybacked and built upon mental computations that had their origins in social interactions.

Are monkeys different from other animals?

What, then, are the differences between monkeys' social knowledge and that of other species? The short answer is that we just do not know. It is now clear that gregarious species like hyenas and pinyon jays recognize other individuals' relative ranks; hyenas may also recognize other individuals' kin. In their basic knowledge of others' ranks and kinship relations, therefore, there may not be many differences between monkeys and nonprimates. However, there are at least five other components of baboons' (and by extension, other monkeys') social knowledge that have not thus far been documented in birds or nonprimate mammals, and which might yet reveal qualitative differences.

First, baboons are able to track short-term changes in the status of transient relationships like sexual consortships (Chapter 6). Although it seems likely that other animals are also capable of monitoring transient social relationships, the necessary experiments have not yet been conducted.

Second, monkeys appear to classify other group members simultaneously according to both their individual attributes (like rank) and their membership in higher-order groups (like matrilineal families) (Chapter 6). The ability to classify others into hierarchical categories allows baboons to take into account other individuals' rank *and* kinship at the same time. Although animals such as hyenas, elephants, and dolphins may live in groups with similarly nested hierarchical structures, we do not yet know whether nonprimate animals make the same simultaneous assessments.

Third, although female baboons group other females into matrilines, or "equivalence classes" (Schusterman and Kastak 1998), they nonetheless recognize the members of a matriline as distinct individuals who are not mutually substitutable. Recall, for example, that when female baboons hear the reconciliatory grunt of their opponent's kin, they change their disposition toward their opponent and that relative but not toward other members of the opponent's matriline (Chapter 6). Reconciliation is specific to the opponent and the relative whose grunt they hear. Future research may (or may not) show that other animals make the same subtle distinctions within members of the same equivalence class.

Fourth, in treating the grunts of their opponent's relative as a proxy for reconciliation with the opponent herself, baboons show that they assume that the grunt is directed *at them* and is causally related to the dispute. They make this causal inference even though they have not interacted with the signaler. Whether or not other animals are capable of making similar indirect causal inferences when evaluating social signals remains to be determined.

Finally, baboons seem almost irresistibly compelled to recognize other individuals' social relationships, even when these social relationships involve the members of another species. Dogs herd sheep and goats, but it is not known whether even breeds like border collies, which have been specifically selected as herders, are as skilled as the baboon Ahla in recognizing the mother-offspring relationships among their charges (Chapter 2). Perhaps the best way to answer these questions will be to rear dogs, hyenas, dolphins, and pinyon jays with goats, and then sit back to see whether they can resist the temptation to reunite kids with their mothers.

EIGHT

Theory of Mind

We cannot perceive the thought of another person at all, we can only infer it from his behavior. **CHARLES DARWIN, 1840:** *OLD AND USELESS NOTES*

In October 1960, during the trial of D.H. Lawrence's publisher on the charge that Lawrence's novel, *Lady Chatterley's Lover,* was obscene, several clergymen testified for the defense. One, Canon Milford, introduced a subtle distinction. He argued that if, during the scenes in question, a reader of *Lady Chatterley* identified with one of the two lovers, that would not be indecent; however, if a reader assumed the perspective of a third party, observing the lovers from behind a tree, that would indeed be obscene. Years later, reflecting on this opinion, the writer Anthony Powell (1983) concluded that Canon Milford's distinction was an important one, recognizing as it did the "important division of the human race between voyeurs and exhibitionists."

Theory of mind and the intentional stance

Regardless of whether we are exhibitionists or voyeurs, our thoughts and conversations are rife with inferences about other individuals' emotions, motives, and beliefs. Depending on these inferences, we may view another person's behavior as deliberate, accidental, ignorant, or devious, and we may, in turn, attempt to influence or alter his beliefs by telling him what we know to be true—or by lying. The ability to attribute mental states like knowledge and ignorance to both oneself and others is to have what Premack and Woodruff (1978) termed a "theory of mind." A theory of mind is a theory because, unlike behavior, mental states are not observable, although they can be used to make predictions about behavior.

Mental states are always *about* some other thing, be it a physical object, another person, or another mental state. Whenever a person thinks, believes, wants, or likes something, he is in what philosophers call an "intentional state"[1] (Dennett 1987b). Mental states are intentional because of their aboutness, and because they are representations about the world they may or may not be true. You can believe that your eccentric aunt is planning to leave her fortune to you, but when you read her will you may be in for a nasty surprise. Mental state attributions are also recursive, because one element "recurs" inside another at different levels of meaning. You believe that your aunt is leaving you money in her will because she mistakenly believes that you will give the money to her cat orphanage.

In thinking about mental state attribution and intentional states in nonverbal animals, the philosopher Daniel Dennett (1987b) has suggested that we begin with the assumption that an animal is an intentional system, capable of mental states like beliefs and desires. But what kinds of beliefs and desires? Here Dennett's different "levels" of intentionality provide us with a number of alternative hypotheses. Using vervet monkey alarm calls as an example, Dennett begins by considering that a vervet might be a zero-order intentional system, with no beliefs and desires at all. A vervet with zero-order intentionality gives an alarm call whenever he sees a leopard because the sight of a leopard triggers this reflexive response. Alternatively, a vervet might have first-order intentionality and give an alarm call because he *believes* that there is a leopard nearby or because he wants other vervets to run into the trees. At this level, the caller has beliefs and desires but he has no conception of his audience's beliefs, nor need he recognize the possible difference between his own beliefs and another's.

Vervets might also be second-, third-, or even higher-order intentional systems, with some understanding about both their own and others' states of mind. A vervet capable of second-order intentionality gives an alarm call because he *wants* others to *believe* that there is a leopard nearby. At even higher levels, both the signaler's and the audience's

1. Note that the use of "intentional" in this philosophical context is quite different from everyday uses of the term, where "intention" is used synonymously with motive, goal, and purpose.

states of mind come into play. At the third level of intentionality, the vervet gives an alarm call because he *wants* others to *believe* that he *wants* them to run into the trees. True linguistic communication, it has been argued, requires at least second-order intentionality on the part of both speaker and listener (Grice 1957; see Chapters 10 and 11 of this volume).

Whether or not baboons or other animals are capable of distinguishing between their own and others' knowledge and perspectives is, as we will see, a contentious issue. Humans, in contrast, are intentional systems almost to a fault. Inferences about our own and others' mental states govern our social interactions and conversations, affecting what we say and how we interpret other people's behavior. In our courts, we use our judgment of intent to decide if a defendant is guilty or innocent, even when his actions are not in doubt. Our impulse to explain events in terms of motives and beliefs is so strong that we are even inclined to ascribe devious and malevolent intentions to machines like cars and computers when they fail to behave as we want. The ornate recursiveness of mental state attributions reaches perhaps its most baroque and tortuous levels in the reflections of adolescent girls, who have no difficulty parsing a sentence like "He thinks that she thinks that he doesn't know that she doesn't really like him." The apotheosis of mental state attribution, however, can be found in the titles of country music songs, where convoluted references to theory of mind are almost mandatory: *Am I going crazy or just out of her mind?*; *I forgot more than you'll ever know;* and *I wish I didn't know now what I didn't know then.*

Some mental state attributions are relatively simple. If we see someone look into a box, his gaze and attention inform us that he is probably looking at something. If someone's back is turned when another person enters the room, we assume that he cannot see, and hence cannot know, who has entered. Attributions like these are often implicit; we make them without deliberate introspection. Other mental state attributions require more explicit thought and planning. Consider Rosalind in Shakespeare's *As You Like It.* In love with Orlando and disguised as the young man Ganymede, she flees in exile to the forest, where she encounters the equally besotted Orlando. When Orlando confesses his love for Rosalind to her (him), she (Ganymede) offers to pretend to be Rosalind so that Orlando can perfect his wooing techniques on her (him). We have no difficulty following such a silly subterfuge. Would a baboon?

In addition to helping us understand others, a theory of mind allows us explicit access to our own mental states. Introspection not only permits the exquisite agony of reliving real and imagined faux pas, but also allows mental "time traveling." We can travel into the past, meandering mentally from room to room in a fruitless attempt to remember where we left our glasses, and we can project ourselves into the future, planning how to avoid an awkward dinner party. The extent to which such "episodic memory" depends on language, and whether animals, too, are capable of imagining themselves in past and future scenarios is a hotly debated issue.

In this and the following chapter, we explore evidence for a theory of mind in baboons and other animals. We devote this chapter to a review of animals' assessments of other individuals' mental states. We consider animals' ability to reflect upon their own mental states in Chapter 9. In addition to drawing on a rather inconclusive set of examples from baboons, we review some experiments that have been conducted on captive monkeys, apes, and other animals. Often, these are more illuminating. We begin with a review of the development of mental state attribution in children—not because we believe that ontogeny recapitulates phylogeny, or because baboons should be regarded as ugly, badly behaved children. Instead, the extensive research on children provides a framework within which to assess the work on animals.

Children's theories of mind

Anyone who has ever played hide-and-seek with a young child knows that two- and three-year-olds are terrible hiders. The compliant adult diligently searches the room, ignoring the legs exposed from under the bed and feigning repeated astonishment at finding the delighted child hiding over and over again in the same place. The child's operational rule seems to be "if I can't see him, I can't be seen." The other individual's visual perspective is not taken into account. Children are also very poor liars. A two-year-old we know once spent a happy naptime industriously painting the bedroom walls, then warned her parent not to enter the room because it was filled with "bad bees and spiders." There was a glimmer of a theory of mind in this ruse, but not enough to appreciate what makes a plausible lie. A more imaginative explanation for a similar transgression was offered by a three-year-old of our acquaintance who asserted that she couldn't possibly have done anything wrong because "I never do anything on purpose."

Given the ubiquity of mental state attribution in normal adult discourse, it is perhaps surprising that young children are so very bad at it. Indeed, before about four years of age children have considerable difficulty recognizing that another person's beliefs and perspectives can be different from their own, and at variance with reality.

In an early experiment investigating the development of mental state attribution, Wimmer and Perner (1983) presented children with scenarios in which they had to predict the behavior of others. The children watched a show in which a puppet, Maxi, put a piece of chocolate into a blue cupboard. Maxi then left the room, and in his absence another puppet removed the chocolate from the blue cupboard and placed it in a green one. The children were then asked where Maxi would look for the chocolate when he returned. Children under four years of age consistently indicated the green cupboard, the cupboard where they knew the chocolate now was. In contrast, older children correctly pointed out that Maxi would believe that the chocolate was still in the blue cupboard. The younger children's errors were not due to faulty memory, because they remembered where Maxi had placed the chocolate. Rather, it seems that children's ability to represent two incompatible beliefs does not become established until around four years of age (see also, e.g., Hogrefe et al. 1986; Astington et al. 1988).

Children do not simply wake up one morning with the ability to recognize that people can hold false beliefs. The ability to attribute mental states to others develops gradually, and there are hints of mental state attribution even in very young infants (Bartsch and Wellman 1995). Infants as young as six months of age attend to their mother's direction of gaze when inferring where to look, and by 18 months of age they can guess both the direction and the location of an adult's focus of gaze even when it falls outside their own visual field (Butterworth and Jarrett 1991). Infants also seem to have a rudimentary concept of other people's goals and intentions. When six- to nine-month-old infants watched a hand reach repeatedly for one of two toys, they looked longer when the hand suddenly reached for a new toy than when the hand continued to reach for the old toy, but using a different motor pattern (Woodward 1998). The infants appeared to impute intent to the actor and were surprised when her interest in one toy switched to the other.

By one year of age, children begin to recognize that gaze has some referential content and that looking is directed *at* something in the environment (Brooks and Meltzoff 2002). One-year-olds seem to understand that words can be mapped onto objects and actions in the world (reviewed by Fisher and Gleitman 2002; Bloom 2003). Crucially, this understanding is accompanied by a form of "social referencing" in which the child uses other people's direction of gaze, gestures, and emotions to appraise unfamiliar objects and situations. The child acts as if she has developed a tacit understanding that gaze and attention are a reflection of underlying knowledge and motivation (Baldwin 1993; Malle et al. 2001; Phillips et al. 2002; Tomasello 2003). As a result, if an adult is looking at a particular object when she says a new word, the child will assume that the word is the name for that object (Chapter 10). At this age, children also begin to use gestures and sounds to recruit adults' attention, both to themselves and to external objects or events. In pointing toward a desired object, they turn to the addressee as if to check that the message has been received, and they repeat and alter sounds or gestures that have been interpreted incorrectly (Golinkoff 1986; Bretherton 1992).

By the age of 18 months, children develop an understanding of people's likes and dislikes, even if these differ from their own. Repacholi and Gopnik (1997) presented 18-month-old children with a scenario in which an actor expressed a strong affection for broccoli. The children were then given a cracker and a piece of broccoli, and the actor said "Give me some." Even though all of the children had previously demonstrated a decided preference for crackers, they consistently handed the broccoli to the actor. This behavior was not solely the result of greed, because the children would also hand the cracker to another actor if the actor had demonstrated a liking for it.

Finally, by the age of two years children begin to distinguish between ignorance and knowledge in others (O'Neill 1996). In one experiment, two-year-olds were shown a new toy that was then placed on a high shelf. The parent either witnessed the event or was absent from the room when the event occurred. When subsequently asking the parent to retrieve the toy, the children were more likely to name the toy or gesture at it when the parent had been absent than when the parent had been present.

One of the most striking features of children's developing theory of mind is their strong motivation to share knowledge and beliefs with others, and to view others as intentional, sentient beings like themselves. Children of one and two years of age *want* to share their experiences and emotions with others. They are strongly motivated not only to use others' attention to learn new words and assess events but also to inform others about their own intentions and emotions. One consequence of young children's sensitivity to shared attention and emotion is their strong capacity for empathy. In one experiment, 18-month-old children were presented with situations in which they could help an unfamiliar adult achieve a goal, such as handing the adult an object that was out of reach. The children clearly recognized the adult's motives and goals and were strongly motivated to help him, even when they derived no benefit from their actions (Warneken and Tomasello 2006; Tomasello et al. 2005). Such eagerness to empathize is, as we will see, less evident in apes.

In summary, an implicit version of a theory of mind emerges at around 18 months of age, aiding young children in their assessments of intentions, desires, and joint attention and guiding their learning of language. A more explicit awareness of beliefs and knowledge, and the role played by mental states in behavior, does not appear until much later, at around four years of age (Frith and Frith 2003). At this age, children not only begin attributing complex mental states to others but also report the reasons for their own and others' knowledge and beliefs. Long before they pass the false belief task, however, children reveal that they view others as intentional beings with goals, motives, likes, dislikes, and even knowledge. Their ability to compare another's perceptual state with their own forms the basis of a social referencing system that is integral to early word learning and the development of social relationships (Bloom and Markson 1998; Tomasello 2003).

These studies, though, beg an obvious question: If a two-year-old can distinguish between an ignorant and a knowledgeable parent, why can't she solve the false belief task? One possible answer is that the traditional false belief task ("Where will Maxi look for the chocolate?") requires children to make an explicit, verbal assessment of another person's beliefs. In contrast, simpler tests examine implicit knowledge that does not require conscious reflection.

Supporting this view, there is some indication that even very young children do recognize false beliefs in others, but they do so implicitly and without being aware of doing so. In a recent experiment, 15-month-old infants were presented with scenarios in which an actor could have either a true or a false belief about the location of a toy. The setup was similar to the one described for Maxi the puppet. In the true belief scenario, both the child and the actor witnessed a toy first being placed into one box and then moved to a new box. In the false belief scenario, the actor left the room before the toy was switched to the new box and so still "believed" that the toy was in the old (now wrong) box. If infants had a rudimentary understanding of the actor's beliefs, they should have been surprised to see the actor reach into the old box in the true belief scenario, but not in the false belief one. But if their responses were guided solely by their own knowledge of where the toy was hidden, they should have been surprised to see the actor reach into the old box in both scenarios. In fact, the infants were more surprised when the actor reached into the old box in the true belief scenario than in the false belief one (Onishi and Baillargeon 2005; see also Garnham and Ruffman 2001). One explanation for these results is that, although they are not explicitly aware of it, even very young children have a tacit understanding of other individuals' likes, knowledge, and beliefs.

The results are also consistent, however, with a simpler interpretation based solely on learned behavioral contingencies (Perner and Ruffman 2005). The children could simply have learned that people tend to search for an object in the place where they last looked at it. According to this interpretation, the children were not surprised when the actor reached into the wrong box in the false belief scenario because she had last looked at the object when it was in that box. In contrast, the children's expectations were violated when the actor reached into the wrong box in the true belief scenario, because she had last looked at it when it was in the other box. If this explanation is correct, very young children have an implicit understanding about other people's likely behavior (and perhaps even attention), but not about their beliefs. As we will see, similar behaviorist interpretations dog almost every experiment and observation that suggests a form of mental state attribution in animals.

Some exasperating anecdotes from baboons

Baboons and other primates often behave as if they attribute mental states to others. The problem is that their behavior can almost always be explained in terms of relatively simple learned behavioral contingencies, without recourse to a theory of mind. This dilemma is especially obvious in the context of apparently "deceptive" behavior.

For example, when a low-ranking female baboon discovers a clutch of bird eggs, she looks furtively around her and then stuffs as many eggs into her mouth as fast as she can. If she is detected she runs away, averting her face from any onlookers. If another baboon catches up to her, she resolutely keeps her mouth firmly closed, even when the other attempts to pry it open. But if a high-ranking female stumbles on a similar trove, she calmly and deliberately eats the eggs in plain view of others. The inescapable impression is that the low-ranking female is trying to conceal her discovery from others in order to avoid theft. For high-ranking animals, such subterfuge is unnecessary.

How should we interpret the low-ranking female's behavior? It certainly *functions* to deceive. But does she understand that if she conceals her find others will remain ignorant? Or has she simply learned that she will be more likely to keep her eggs if she crams them as quickly as possible into her mouth, out of sight of others?

Similar sorts of apparent deception occur during social interactions. Hannah, the seventh-ranking female at the time, had been receiving attention all morning from higher-ranking females who wanted to handle her baby. Although these females had always been friendly, their constant attentions had prevented Hannah from eating or resting. Hannah had just sat down to eat a fig when Sierra, the third-ranking female, approached and reached for her baby. Hannah grabbed Sierra's hand and cuffed her on the face. Although Hannah's threat violated the established rank order, Sierra did not retaliate but moved away. An hour later, Sierra approached Hannah again. Perhaps remembering that she had hit Sierra earlier, Hannah flinched and began to move away, but she relaxed when Sierra began to grunt. As soon as Sierra reached Hannah, she leapt on her and bit her on the neck.

How do we explain this anecdote? A rich explanation, based on the assumption that baboons possess a full theory of mind, argues that Sierra had been stewing angrily about Hannah's behavior for the entire hour and wanted to exact revenge. Her grunts were intended to deceive Hannah into thinking that Sierra was no longer angry and only wanted to handle her baby. The simpler explanation makes no recourse to mental state attribution. It argues only that Sierra approached Hannah in order to handle her baby and then remembered that Hannah had threatened her earlier. She therefore bit Hannah in delayed retaliation.

The problem is that both explanations are, at least in principle, equally plausible. Here is another example. When a high-ranking, potentially infanticidal male begins to display and chase other males and females, mothers with infants often hide to avoid detection. Often, their hiding attempts are rather feeble. The female crouches against a tree trunk, so that her face is hidden but much of her body is not. Her sorry attempt at concealment recalls young children's attempts at hide-and-seek. Nonetheless, some females are excellent hiders. Amelia, for example, always concealed herself in the densest clump of grass or the thickest and most impenetrable thorn bush. Although very low-ranking, she never lost an infant to infanticide. Had Amelia simply learned that she was unlikely to be chased if she crouched in dense bushes, or was she aware, at some level, of what other individuals could or could not see?

In other contexts, there is little ambiguity about the perplexing failure of baboons to recognize the perspectives of others. When baboons cross from one island to another at the peak of the flood, they typically choose the shortest and shallowest route. If the flood is large, however, they are often forced to wade or swim for hundreds of meters (Chapter 3). Young infants are carried ventrally and can be completely submerged for several minutes. Several years before we began our study, we were told, Sylvia's baby had drowned on a long crossing. Sylvia was apparently oblivious to the plight of her infant and never attempted to place him on her back. She acted as if she believed that, as long as *her* head was above water, everyone else's head must be above water too.

Perhaps it is unreasonable to expect baboons to understand the relation between breathing and life. But baboons often also seem to be insensitive to small juveniles' intense fear and distress during long water crossings. At the initiation of any water crossing, the smaller juveniles congregate at the water's edge, whining, screaming, writhing on the ground, and lashing their tails (Fig. 31). Some mothers backtrack to retrieve their offspring, or wait at the other side of the crossing for their offspring to emerge. Similarly, some adult males will occasionally carry the juvenile offspring of their former friends on their backs. Other mothers, though, forge blithely ahead, leaving their offspring to fend for themselves. There have been several instances when young juveniles have failed to make the crossing, either because a predator killed them

Figure 31. Two young juveniles huddle together anxiously as they watch the group initiating a water crossing. Photograph by Anne Engh.

after the rest of the group had left, or because they drowned as they struggled to make the long crossing on their own. Although the mother of an abandoned juvenile sometimes gives contact barks in response to her offspring's panicked screams on the opposite shore, she rarely crosses back to retrieve it. And eventually she moves away with the rest of the group—lagging behind, looking back at the crossing, and giving occasional contact barks as the screams fade into the distance.

These Dickensian observations are puzzling from a number of perspectives. First, even the apparently callous mother who abandons her offspring at long water crossings is not entirely indifferent to her offspring's plight; she appears genuinely concerned by its agitated screams. But she seems to fail to understand the cause of this agitation. She behaves as if she assumes that if she can make the water crossing, everyone can make the water crossing. Other perspectives cannot be entertained. Second, when a dependent infant dies, mothers often carry the body for several days and give loud contact barks for several more days after they finally abandon it. Why would a mother who gives contact barks for days after the death of an infant walk away from a stranded young juvenile? The easy, proximate explanation is that hormones such as oxytocin, which are associated with lactation, cause mothers to seek continuous contact with their infant, and that this motivation disappears after weaning. Recall, though, that females who lose a close relative to predation or a sudden disappearance experience elevated glucocorticoid levels, a clear indication of stress (Chapter 5). Thus mothers respond physiologically, even though they do little to prevent their juveniles' loss.

Suppose baboons had a full-blown theory of mind. Would this make their lives any different? We humans are naturally inclined to make use of our theory of mind when negotiating complex social relationships. But a simpler strategy based on the memory of past interactions can be almost as successful and result in ostensibly similar outcomes. People with Asperger's syndrome, a highly functional form of autism, often report that they have difficulty recognizing complex mental states like envy and love in others. Rather than relying on their (often faulty) assessment of others' mental states when a complex social interaction presents itself, they depend on a "library" of carefully stored memories of previous interactions when deciding how to respond. At the same time, though, people with Asperger's syndrome are also aware of the strategy's limitations and the computational costs of calculating, through learned contingencies, what others are feeling and thinking. Temple Grandin, an autistic professor of animal behavior, has argued that animals' responses to social events are likewise guided by memories of previous interactions rather than theories about mental states (Grandin and Johnson 2005). Instead of analyzing beliefs and desires, animals use past performance as a guide to future behavior.

Indeed, there is almost no example of possible mental state attribution in animals that cannot also be interpreted in terms of contingency learning. People often claim that their dog feels guilty when he does something that he "knows" he shouldn't do, like knock over the garbage container. Dogs certainly act as if they know they have done something wrong; their Tai Chi slink is a caricature of guilt. But dogs often assume the same guilty postures if another dog, or even a human, tips over the garbage. Rather than experiencing guilt or remorse about their actions, dogs are probably responding on the basis of a learned contingency: when garbage is on the floor, people begin to shout (Grandin and Johnson 2005).

The neural correlates of a theory of mind

The anecdotes from baboons suggest that monkeys, like young children, fail to understand other individuals' beliefs, perspectives, and predicaments. However, the evidence from young children also suggests that implicit recognition of simple mental states—like intentions and emotions—emerges at a considerably younger age than more explicit attributions of knowledge and beliefs. This discontinuity may occur in part because belief attributions are associated with different brain regions from those involved with representing goal-directed action.

A thorough review of the neural correlates of "theory of mind" is beyond the scope of both this book and our expertise (for reviews see Frith and Frith 2003; Gallagher and Frith 2003; Saxe et al. 2004). What we wish to emphasize here is recent neurological evidence that suggests that monkeys, like young children, might be capable of simple forms of mental state attribution even if they do not pass false belief tests. When observing other animals' gaze and actions, monkeys and humans show a similar pattern of neural activity. When attributing knowledge, belief, and other mental states, however, humans' brains also show activation in areas that have undergone enlargement recently in our evolutionary history, since the ancestral line leading to humans diverged from the common ancestor of humans, monkeys, and apes.

As noted in Chapter 7, in both monkeys and humans the perception of gaze direction and goal-directed behavior appear to activate the same relatively primitive areas of the brain, including the superior temporal sulcus (STS) and the amygdala. The STS is particularly sensitive to the orientation of another individual's eyes (Jellema et al. 2000; Emery and Perrett 2000; Chapter 7). Mutual gaze evokes greater activity in the STS than does averted gaze, suggesting that the STS facilitates the processing of social information (Pelphrey et al. 2004). In both monkeys and humans, STS also responds to goal-directed actions and perceptions. Cells in monkeys' STS show particular increased activity to goal-directed hand movement when the actor they are observing is gazing at his hand (Jellema et al. 2000; Lorincz et al. 2005). It therefore seems possible that STS may be involved in representing what others see and what their actions and intentions are (Gallagher and Frith 2003). Similarly, in both monkeys and humans the amygdala responds strongly to social stimuli, particularly aversive ones. It also seems to be important for processing information about gaze direction (Adolphs et al. 1999; Kawashima et al. 1999; Fine et al. 2001; Santos et al. 2006a).

Other areas of monkeys' brains seem to be sensitive to the intentions that underlie behavior. As in humans, "mirror neurons" in the inferior parietal lobule (IPL) of monkeys' brains are activated both when a monkey performs a specific action and when he observes someone else performing that action. Furthermore, neurons that code for specific acts, such as grasping, seem to be context-dependent. Some mirror neurons in the monkey respond more when the monkey grasps a piece of food to eat it than when he grasps the same food to place it into a container. This same context-dependence is preserved when the monkey observes *another* individual perform these actions. Significantly, many neurons begin to fire *before* the other individual actually performs a specific action—that is, before he grasps-to-eat as opposed to grasps-to-place. Thus, it seems possible that these neurons encode not only the specific motor act but also the actor's intentions (Fogassi et al. 2005; see also Nakahara and Miyashita 2005; Rizzolatti and Craighero 2004; Rizzolatti and Buccino 2005). These results are perhaps not surprising, given the benefits of being able to predict what others are going to do. A monkey should not risk a fight over a piece of food that another monkey is grasping to eat.

In adult humans, however, tasks associated with belief attribution also activate other areas of the brain, including the temporal lobes, the temporo-parietal junction, and the medial prefrontal cortex (Apperly et al. 2005). One part of the prefrontal cortex, the anterior paracingulate cortex (ACC), shows particular activation in tasks involving the representation of mental states. This area of the brain has undergone comparatively recent evolution in both size and complexity (see Saxe and Powell 2006 for further discussion about other brain regions activated when reasoning about other individuals' thoughts). It also develops very slowly during childhood (Fuster 1997; Frith and Frith 2003; Saxe et al. 2004; Rilling 2006). Some projection neurons in the ACC are present in apes and humans but not in monkeys and are more developed in humans than in apes (Frith and Frith 2003). Moreover, in humans they are not present at birth but begin to appear only at around four months of age. Unfortunately, it has not yet proved possible to apply functional neuroimaging techniques to infants and young children, so we can still only speculate about the neural mechanisms underlying the developing theory of mind.

In summary, the evolution of a full-blown theory of mind in humans is correlated with the enlargement of specific areas of the brain, especially the prefrontal cortex. Other areas of the brain, including the STS, help to interpret behavior and intentions and aid in imbuing content and context to thought (Frith and Frith 2003), but they do not appear to be involved in representing knowledge and beliefs. Taken together, therefore, current neuroanatomical evidence predicts that monkeys and apes, like young children, might be able to represent simple mental states like emotions and intentions even if they cannot recognize more complex ones like knowledge and beliefs. We explore this hypothesis in more detail below.

Recognizing ignorance and knowledge in others

When moving through wooded areas or when separated from others, baboons often give loud, clear barks that can be heard for over 500 meters. Because they often appear to be exchanged between widely separated individuals or subgroups, the barks seem to function as "contact" calls (Fig. 32). Contact barks grade acoustically into alarm barks, but as we describe in Chapter 10, baboons respond to these vocalizations as if they were discretely different call types. Analogous "contact" calls occur in many other species of monkeys and apes. Male chimpanzees, for example, often give loud "pant hoots" upon arriving at a food resource (Wrangham 1977; Clark and Wrangham 1994). Typically, more calls are given at large food patches than at small ones.

Despite these observations, there is some doubt about whether loud calls have evolved specifically to alert others to food or to maintain contact between separated individuals. Capuchin monkeys also give a specific call when they find a desirable food item. Rather than functioning to recruit others, though, these calls seem to function to announce pos-

Figure 32. Baboons who have become separated from the group often give contact barks. Photograph by Anne Engh.

session and to warn listeners that the owner will defend her prize (Gros-Louis 2004). The calls certainly *function* as "food" calls—a listener can easily learn to associate these particular calls with the discovery of food. At a proximate level, however, the caller is attempting to deter, not attract, others. Similarly, although chimpanzees give more pant hoots at large food patches than at small ones, they do not call more when they discover a new food source than when they are revisiting an old one (Clark and Wrangham 1994).

Calls like these highlight a problem common to all studies of long-range "contact" or "food" calls: although listeners can use the calls to maintain contact or locate food resources, signalers may not call with the purpose of informing ignorant animals of the group's location. For anyone interested in the existence (or lack) of a theory of mind in animals, the mechanisms underlying the production and perception of these calls are crucially important. In the case of baboons, it seems clear that individuals give contact barks because they have lost sight of others and are feeling anxious. What is less clear is what listeners think when they hear the contact barks of others. The key to solving this puzzle is: who answers?

An individual who attributes mental states to others recognizes that her knowledge may be different from someone else's. A baboon who possesses a theory of mind, therefore, realizes that other individuals give contact barks when they are separated from the group and want to know where the rest of the group is. If she understands that another group member is lost (and if she is motivated to help), she should answer even when she herself is at the center of the group and at no risk of becoming lost. On the other hand, if baboons do not realize that other individuals' mental states can be different from their own, they should answer only when they themselves are also at risk of losing contact with others. In this case, whether or not a baboon produces a contact call will be determined primarily by her own circumstances (separated or not) rather than her perception of another's.

To test between these hypotheses, we designed an experiment in which an adult female baboon was played the contact bark of a close adult relative whom she might be motivated to answer—either a sister, mother, or daughter. Eighteen females served as subjects. We conducted trials when the group was moving through thickly wooded areas, and we timed our trials so that there was variation in the subject's position in the group and in how far she was from others. All trials were conducted during a time when no group member had given a contact bark within the last half hour. In response to our playbacks, subjects oriented in the direction of the speaker in 20 of 36 trials (56%), and "answered" their relative's contact bark in seven trials (19%). No unrelated females ever answered the contact barks.

At first inspection, this rate of "answering" might be taken as weak evidence for the selective exchange of contact barks among close kin. But closer examination suggested otherwise. Females answered their relatives' contact bark primarily when they themselves were lagging behind the group and separated from other females. Subjects who were in the last third of the group progression were significantly more likely to answer their relative than those in the first two-thirds. Similarly, subjects were more likely to answer when they had no other female within 25 meters than when there was at least one other female nearby. In short, female baboons answered contact barks primarily when they themselves were separated from the group. They answered their relative's contact bark because hearing the call made them look around and realize that they had become separated from a close companion. It is as if they were thinking "That's Shashe calling. Hey, I'd really like to be with Shashe now," rather than "Shashe's lost; I'll let her know where I am" (Cheney et al. 1996).

We subsequently replicated these results by playing infants' contact barks to their mothers and to unrelated females, reasoning that mothers should be particularly concerned about their infant's whereabouts and very responsive to signals indicating that their infant had become separated from them. Infants are highly vulnerable to infanticide and predation, so there ought to be strong selective pressure for mothers to answer separated infants' contact barks, even when they themselves are at no risk of becoming lost.

Baboons apparently do not appreciate this logic. Mothers clearly recognized their infants' contact barks, because they were significantly more likely to orient and move toward their infants' calls than toward the calls of other infants. Despite their strong behavioral responses, however, mothers were no more likely to answer their own infant's calls than they were the calls of another infant. Instead, whether a female answered depended again on her own position in the group. If a female was separated from others, she was likely to answer the infant's call—regardless of whether or not it was her own infant. If she was not alone, she rarely answered any infant's call (Rendall et al. 2000).

An individual who calls to maintain contact with separated group members should answer the calls of others even when she is in the center of the group progression. The ability to answer others under these conditions, however, requires that a baboon recognize that others can be lost even when she herself is not. The fact that baboons do not selectively answer others' calls suggests that they do not give contact barks in order to provide others with information. The calls *function* as contact calls because they allow a listener to deduce other individuals' locations. When a group is moving through a wooded area, there are likely to be several individuals who find themselves out of sight of kin or other close companions at the same time. Through their calls, separated individuals can remain in contact until they locate each other and the rest of the group. But communication is, in a sense, inadvertent, because the listener extracts information from a signaler who has not, in the human sense, intended to provide it.

These results are consistent with many other observations of birds and mammals suggesting that signalers do not recognize whether or not their audience is ignorant or already knowledgeable about the information being conveyed. Vervet monkeys and baboons, for example, continue to give alarm calls long after all group members have been alerted to the presence of a predator. Similarly, experiments conducted with captive rhesus and Japanese macaques suggest that mothers do not recognize when their offspring is ignorant or already informed about the presence of a nearby predator; they give alarm calls at the same rate in both contexts (Cheney and Seyfarth 1990).

Perhaps because they cannot attribute mental states different from their own to others, monkeys and apes also seem not to *expect* to be informed or deceived. For instance, chimpanzees who remain silent when they find a fruiting tree are not punished by those who arrive later (Clark and Wrangham 1994). Similarly, when baboon mothers lose track of their infants, they sometimes become quite frantic in their attempts to locate them, climbing into trees and giving contact barks for long periods of time. But when the infant finally leaves his play group and reappears, his mother never cuffs or admonishes him for ignoring her. To do so would require that she infer the intent to ignore. Instead, the mother simply stops giving contact barks.

Baboons' apparent insensitivity to others' plight is especially puzzling when a predator attack or a deep water crossing splits the group. As we described in Chapter 3, lion attacks sometimes divide the group into widely scattered subgroups for several days. These separations are clearly stressful, because females' glucocorticoid levels are significantly higher after lion attacks that cause a group split than after attacks that do not (Engh et al. 2006b). During these periods of separation, members of different subgroups give contact barks at a high rate and respond to the contact barks of other subgroups. When the group finally reunites, however, the separated parties do not run up to each other in joyful embrace. They simply approach each other, cease calling, and resume foraging. It is as if, once reunited, all emotions surrounding the separation simply disappear.

Equally enigmatic was the Lord of the Flies incident in 2004. The annual flood was especially large that year, and the baboons had to swim from island to island—something that the juveniles were loath to do. One day, the adults crossed to another island, leaving almost all of the group's juveniles stranded behind. (Interestingly, within a day the juveniles' ranks had become completely size-dependent. The young offspring of high-ranking females who had previously been able to supplant juvenile females twice their size fell to the bottom of the hierarchy, while the older offspring of low-ranking females became dominant.) Not surprisingly, the juveniles seemed to find this separation very distressing. They foraged around a small section of the island in a close, bereft unit, and at night slept tightly bunched in the same tree (Fig. 33). Throughout the day and night, they gave agitated contact barks and screams. Those of us who were observing the adults on the next island could hear the juveniles' calls. The adult baboons, too, appeared to hear

Figure 33. When the juveniles were left behind on an island, they foraged together in a tight, bereft bunch. Photograph by Anne Engh.

their offspring calling, because they often oriented toward them. But only one adult ever produced an answering bark. This bark, as well as a distant copulation call, produced a flurry of frantic calls from the juveniles—but these were never answered. The juveniles and adults were reunited three days later. But the adults did not come to the juveniles; instead, the juveniles swam to the adults.

These results again suggest that baboons do not produce calls in response to their perception of another individual's ignorance, predicament, or ability. From a human's perspective, it is as if a mother were chatting blithely with her friend in a busy supermarket, ignoring her toddler's cries in the next aisle. On the other hand, it is always possible to argue (if perhaps implausibly) that baboons are behaving adaptively even if they fail to respond to each other's calls. As long as she knows where her infant is, there is no real urgency for a female baboon to run to collect it or to answer its calls. As long as the adult baboons knew where the juveniles were during the Lord of the Flies separation, there was no immediate need to reunite.

But once again, the situation is more complicated than this, because for baboons it is not simply a case of "out of sight, completely out of mind." During the same flood of 2004, a two-year-old male orphan, Harley, became stranded alone after the group swam to another island. Normally, a lone juvenile isolated under these circumstances would have been killed by a predator. Harley, however, was resourceful. First he joined a group of impala and foraged with them for two days. Then, perhaps tiring of the impalas' rather tedious company, he joined a group of vervets, who tried but failed to chase him away. Throughout his separation, Harley gave contact barks that were occasionally "answered" by his aunts on distant islands. But the group never came to Harley's island, and Harley was left to forage alone with the vervets. Finally, five weeks later, the group came to an island that was separated from Harley's island by only 50 meters of water. The group gathered at the water's edge, grunting and barking at Harley. Whenever Harley approached the water, the group responded with a chorus of loud, rapid grunts. Finally, after several forays, Harley swam to other side. As soon as he emerged, his aunts ran up to him and attempted to groom him. First, however, Harley approached a male who had immigrated into the group during his separation and presented to him. The other group members obviously recognized Harley, and treated his reunification as more than a casual event. Indeed, their behavior during this incident makes their diffidence during briefer but more stressful separations even more difficult to understand.

A variety of results argue, therefore, that baboons and other primates do not produce vocalizations in response to their perception of another individual's ignorance or circumstance. They appear not to understand that their own knowledge and abilities might be different from someone else's. But might they be capable of recognizing other individuals' intentions and motivations?

Recognizing the link between seeing and knowing

Cooperative contexts

A colleague of ours, Conrad Brain, was once observing baboons in a desert canyon in Namibia when he was approached by a juvenile, Chloe. Chloe sat down several feet in front of him and gazed into his eyes. Next, she looked down at the crevice below the rock where Conrad was sitting and gave a soft alarm bark. She then gazed into his eyes again and repeated the alarm call. Conrad followed Chloe's direction of gaze and saw a spitting cobra lying just below his feet. Once he had moved away, and taken time to think about the incident, what struck Conrad most forcefully was his strong impression that Chloe had deliberately made eye contact with him before she gave her alarm call.

Many studies have shown that monkeys and apes attend to other individuals' eyes and their direction of gaze (Perrett and Emery 1994; Emery 2000; Deaner and Platt 2003; Ghazanfar and Santos 2004). They use gaze to target opponents and to recruit support in aggressive alliances (Sato and Nakamura 2001; Tomasello et al. 2001). Monkeys also seem to recognize that displays and facial expressions are ineffective without some degree of eye contact. They rarely display at another individual if that individual's back is turned to them, and they make concerted efforts to engage eye contact before initiating a friendly or aggressive interaction. Perhaps for this reason, baboons and other monkeys also often go to some lengths to *avoid* eye contact when someone begins to threaten them. For example, when an adolescent male baboon begins to threaten a female, the female will often take an intense and sudden interest in a particular part of an adult male's back and begin to groom it assiduously. This drives the adolescent male into paroxysms of frustration. He circles the female, trying to gain eye contact with her, while the female slowly adjusts her position to ensure that her back is always turned toward him. The adolescent male lunges toward the female; he threat-grunts; he leaps up and down; he thrashes branches. All the while, the female continues her intense grooming, seemingly oblivious to the commotion surrounding her. (The encounter usually ends when the adult male cracks and chases the adolescent away.) The female's attitude might be summed up as "if I'm not looking at you, you can't really threaten me."

Monkeys and apes also seem to recognize the relationship between an individual's gaze and the target of his attention. If a chimpanzee observes a human gazing at something behind a barrier, she will search behind the barrier, as if she recognizes that gaze is always directed *at* something or someone (Tomasello et al. 1999). Similarly, in one experiment conducted with Diana monkeys (*Cercopithecus diana*), a subject was shown a photograph of a familiar cage-mate looking in a particular direction. If a toy then appeared in a location opposite from the one in which the monkey in the photograph was looking, subjects reinspected the photograph, suggesting a violation of expectation (Scerif et al. 2004).

Other experiments, however, suggest that monkeys and apes may not always understand that they can use another individual's direction of gaze to gain information about objects or events in the environment.

In a test replicated many times and in many forms, a human presents a chimpanzee with a choice of two boxes. One box contains food, the other is empty. The experimenter then looks, gestures, or points to the box containing the food. Although chimpanzees can eventually learn to use these cues to locate food, they do not do so on the first trial; their choices are essentially random (e.g., Tomasello et al. 1997; Povinelli et al. 1999; Call et al. 2000; Bräuer et al. 2006). Furthermore, having learned the contingency rule "choose the box at which the human is looking," they do not readily transfer this information to other gestures, such as pointing. Similar negative results have been obtained with monkeys (e.g., Anderson et al. 1996; Vick et al. 2001; Neiworth et al. 2002).

Another series of experiments using somewhat different techniques also concluded that chimpanzees fail to understand that seeing, attention, and knowledge are tightly linked. Chimpanzees who knew how to beg for food by gesturing at a human were presented with two potential human donors. One person looked at the chimpanzee. The other either looked away, had her eyes blindfolded, or had a screen in front of her face (or any of many other variations on this theme). If chimpanzees understood the link between gaze and attention, they should have gestured only at the person who could see them. However, they were as likely to gesture at the person who could not see them as the one who could (Povinelli and Eddy 1996).

This is not to say that this sort of task is beyond the capacity of any animal. In particular, domestic dogs are very adept at using gaze or gestures to determine which of two locations has food. When presented with a human or another dog informant, they reliably choose the location where the informant is looking, pointing, or orienting (e.g., Hare et al. 1998; Hare and Tomasello 1999; Miklosi and Topal 2004). Indeed, in one direct comparative experiment dogs were considerably more accurate than chimpanzees in their ability to use communicative cues like pointing, gazing, and reaching to locate food (Bräuer et al. 2006). In addition to using other individuals' direction of gaze to gain information, dogs often go out of their way to make eye contact with others before attempting to communicate with them, and they appear to be sensitive to whether a person is attentive or inattentive (Gacsi et al. 2004). When a dog wants to play ball, he will often drop the ball at the deserving human's feet and gaze back and forth from the human's eyes to the ball (often supplementing his pleas with a few whines). If the human is in a sadistic mood and averts his head, the dog may circle around the human until he is facing him once more and then repeat the process.

Conrad Brain's encounter with Chloe and the cobra should come as little surprise to anyone who has ever lived with a ball-obsessed dog.

Given chimpanzees' larger brains, the superior skill of dogs at recognizing the informative content of gaze and gestures is puzzling. Some scientists attribute the difference to the different social and ecological environments in which primates and social carnivores evolved (Hare and Tomasello 1999; Hare et al. 2002). Dogs are descendants of wolves, carnivores that breed and hunt cooperatively. When coordinating their movements during cooperative hunts, it is essential to monitor other individuals' gaze cues and orientation. Dogs have also been artificially selected for thousands of generations to be acutely sensitive to human behavior and attention. In contrast, under natural conditions monkeys and chimpanzees do not routinely share food or inform each other of the location of food, nor does their survival depend on coordinated group hunts. Further, monkeys and chimpanzees never naturally point or gesture to one another while foraging. For this reason, they may not readily attend to such cues when interacting with human experimenters.

Competitive contexts

These observations led Brian Hare and his colleagues (2000) to predict that chimpanzees might show greater aptitude for recognizing the relation between seeing and knowing if they were tested in more competitive contexts. In a series of ingenious experiments, they designed a testing arena in which a subordinate chimpanzee had to compete against a more dominant group-mate to acquire food. The dominant chimpanzee sat in a holding area at one end of the arena and the subordinate at the other end. Food was then placed in the center of the arena. One piece of food was visible to both chimpanzees. The other piece of food was visible only to the subordinate, because it was hidden from the dominant by a barrier. The experimenters reasoned that, if the subordinate chimpanzee had some understanding about the visual experiences of others, she should expect the dominant chimpanzee to approach the food that she could see. The subordinate should therefore take advantage of this by snatching the food that the dominant could not see (Fig. 34). Indeed, the subordinate chimpanzee most often (but not always) approached the food that the dominant could not see (Hare et al. 2000). The results of these and subsequent experiments led Hare and his colleagues to conclude that chimpanzees do in fact understand what others can and cannot see, but that this ability is evident only in competitive contexts.

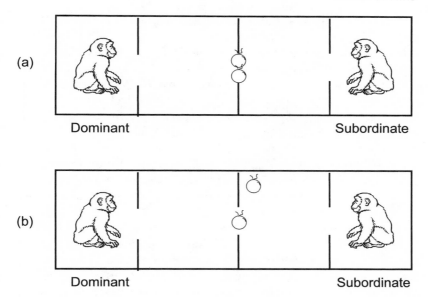

Figure 34. The protocol for one of Hare et al.'s experiments. In one condition (a), both pieces of food were visible to both chimpanzees. In the other condition (b), one piece of food was visible to both chimpanzees, while the other was visible only to the subordinate. Subordinates were significantly more likely to obtain food in condition (b), and they were most likely to approach the piece of food that was visible only to themselves. After Hare et al. 2000.

In fact, in subsequent experiments that deliberately compared chimpanzees' performance in two cognitive tasks (an object choice task and a discrimination task) when competing with a human or cooperating with him, chimpanzees performed better in the competitive context (Hare et al. 2001; Hare and Tomasello 2004).

Hare and his colleagues' claims have proven controversial, in part because they have not been easy to replicate in tests with other chimpanzees (Karin-D'Arcy and Povinelli 2002). They also beg the question of why chimpanzees' ability to understand what others can and cannot see should be so fragile and so difficult to extend to less competitive contexts. In fact, there is mounting evidence that chimpanzees' performance is strongly affected by the procedural methods of the experiments (Barth et al. 2005). We return to this vexing question later in this chapter.

Results similar to Hare's have emerged in experiments conducted with monkeys. Working on Cayo Santiago, an island off the coast of Puerto Rico that is home to hundreds of rhesus macaques brought from India in the 1930s, Flombaum and Santos (2005) presented individual monkeys with a choice of two grapes, each placed on a small platform in front of a person. One person looked at the monkey, while the other averted his eyes, closed his eyes, turned his back, placed a barrier in front of his eyes, and so on. The monkey's job was to approach and "steal" the grape from one of the two people. Note that the design of this experiment was very similar to the one in which chimpanzees tried to beg food from a seeing or an unseeing person, except that in this case the context was competitive rather than cooperative. Almost all of the monkeys chose to steal the grape from the person whose gaze was in some way averted. These results led Flombaum and Santos to conclude that rhesus macaques possess a crucial component of rudimentary theory of mind: the ability to reason about what other individuals can see on the basis of where they are looking.

In a subsequent experiment, individual monkeys had the choice of "stealing" one of two transparent boxes that each contained a grape. One box was covered with noisy bells; the other was not. The boxes were placed on the ground in front of a human. In some trials, the person looked down, with his head between his knees, so that he could not see the boxes. In others, he looked straight ahead, so that he could see both the boxes and the monkey. In trials when the person was looking down, monkeys consistently approached the silent box. But when the person was looking at the monkey, subjects were as likely to choose the noisy box as the silent one (Santos et al. 2006b). The authors conclude that monkeys know both what others can and cannot hear and what they can and cannot see. They therefore understand that there is no point in being silent and stealthy when a person can see them. However, the fact that monkeys were willing to approach the boxes at all when the person was looking at them (in marked contrast to the experiment described in the previous paragraph) casts some doubt on the assumption that the monkeys viewed the context as a competitive one that required deceptive behavior. Additional experiments will be required to test this hypothesis.

There is some support, then, for the socioecological hypothesis that monkeys and apes are more sensitive to other individuals' visual perspective in competitive, as opposed to cooperative, contexts. Santos and colleagues (2006a) speculate that an underlying cause for this discrepancy may be the role played by the amygdala in lending salience and emotional content to events. As mentioned earlier, the amygdala is activated by salient social stimuli, especially those involving negative emotions or experiences (Winston et al. 2002). It also seems to play an important role in reasoning about eye gaze and orientation (Leonard et al. 1985; Kawashima et al. 1999; Fine et al. 2001). As a result, monkeys and apes may be more attentive to other individuals' gaze and intentions in competitive than cooperative social situations.

Nonprimate animals

The experiments just described support the suggestion that monkeys and apes recognize the link between seeing and knowing—at least in some contexts—and may therefore possess a rudimentary theory of mind. If we accept this explanation for monkeys and apes, however, we should also extend it to dogs, because dogs outperform chimpanzees in their ability to use communicative cues like pointing, gazing, and reaching to find food. We might even extend the explanation to birds.

Many members of the corvid (crow) family—especially jays, ravens, and nutcrackers—cache food for later retrieval. They also pilfer from the caches of others. Perhaps as a result, individuals take great pains to conceal their caching sessions from others. If another bird is present when a bird caches food, the cacher will choose sites that are distant, out of view, or hidden behind barriers. If they sense that they are being observed, cachers will also interrupt a caching session to change sites or recover food from previously hidden sites (Emery and Clayton 2001; Bugnyar and Kotrschal 2002, 2004; Emery et al. 2004; Dally et al. 2005). These behaviors are much less likely to occur if a bird is allowed to cache his food in private.

Potential pilferers, in turn, often watch inconspicuously and at some distance (Bugnyar and Kotrschal 2004), acting as if they are deliberately withholding their intentions. In one experiment conducted by Bugnyar and Heinrich (2005), ravens (*Corvus corax*) were given the opportunity to cache food in the presence of two other ravens (Fig. 35). Although the cacher could see both of the other birds, only one of these birds (the "knowledgeable" observer) could see the caching arena; the view of the other bird (the "ignorant" raven) was blocked by a barrier. When each of these potential pilferers was later released into the arena, the cacher recovered food more rapidly in the presence of the knowledgeable than the ignorant raven. The authors argue that the cacher's behavior could not be explained by the simple contingency rule "recover food if there was another raven present during caching," because the cacher had been able to see both the knowledgeable and the ignorant bird during his caching session. Instead, cachers behaved as if they recognized that another bird's success at pilfering food depended on his prior visual perspective. Whether they had learned about this association through

Figure 35. While an observer scrutinizes him, a raven decides where to hide his food. Photograph by Thomas Bugnyar.

experience or whether they had some understanding of the relation between seeing and knowing is not clear. Similarly, scrub jays are more likely to re-cache food when they are in the presence of a bird who observed their initial caches than when they are with a bird who did not (Dally et al. 2006)

Inferring knowledge or inferring intent?

The evasive ploys used by caching and pilfering ravens are functionally deceptive, in the sense that they manipulate competitors' knowledge. In a similar way, when rhesus macaques attempt to steal a grape from an unseeing observer, they appear to be acting deliberately to avoid detection. Similar apparently deceptive ploys have been described for many other species of animals, including chimpanzees, monkeys, and even dogs (see examples in Byrne and Whiten 1988; Cheney and Seyfarth 1990). We once knew a dog in central London, Nicky, who was forbidden to enter his owner's living room because he knocked over valuable pieces of art when he wagged his tail. One night, Nicky's owner looked up from his newspaper to discover Nicky lying at his feet. Wondering how he had managed to enter the room undetected, the owner ordered Nicky out of the room. Then, peeking around his newspaper, he watched as Nicky returned, laboriously inching across the floor on his heels. The owner then realized that he was usually alerted to Nicky's entrance by the noise his toenails made as they clicked across the wood floor. Apparently, Nicky had learned that he could avoid detection if he entered silently, by walking on his heels.

How had he reached this conclusion? An explanation based on theory of mind would argue that Nicky had learned that his owner would remain ignorant if he (Nicky) remained silent. A more parsimonious explanation would posit simply that Nicky had learned by trial and error that, for some baffling reason, his owner never shouted at him if he entered the room using a silly walk. (In any case, the owner decided to move the pieces of art.)

In a similar way, a raven that skulks nonchalantly in the background while his rival hides food might recognize that if the cacher cannot see him he cannot know that he is being observed. Alternatively, the raider might simply have learned that he is more likely to be able to steal another raven's cache if he remains at some distance from the cacher while the latter hides his food.

More generous interpretations that bestow a degree of theory of mind to animals, then, can usually be brought abruptly down to earth by stingier behaviorist explanations based on learned associations. This is not to say that these stingier explanations are completely correct, or that we should reject the argument that animals are capable of recognizing at least some components of other individuals' mental states. Consider the experiments suggesting that chimpanzees and rhesus macaques know what others can and cannot see. The results are certainly consistent with this interpretation, and they are difficult to explain in terms of learned contingencies. But they are also consistent with a simpler mentalist interpretation that posits that animals use gaze direction to assess not other individuals' knowledge, but rather their intentions and motivations. According to this interpretation, animals may not understand the relationship between seeing and knowing. They do, however, understand the relationship between seeing and *intending*. As a result, they recognize, for example, that other individuals are motivated to defend food that they are looking at, and less likely to defend food in which they show no interest.

This middle ground interpretation may help us resolve the apparent contradiction between the wily grape-stealing macaques of Cayo Santiago—whose behavior suggests an understanding of other individuals' knowledge—and the oblivious contact-barking baboons of the Okavango—whose behavior suggests they do not. It may also help to resolve the many other experiments and anecdotes suggesting that monkeys fail to understand other individuals' visual perspectives, even in competitive contexts (Kummer et al. 1996).

Recently, Burkart and Heschl (in press) replicated Hare et al.'s (2000) experiments with common marmosets (*Callithrix jacchus*), a species of New World monkey not noted for its especially large brain. They presented marmosets with a choice of two food items, one visible only to themselves and the other visible also to their dominant rival. Subordinate marmosets, like subordinate chimpanzees, consistently chose the food visible only to themselves. These results suggested that marmosets, like chimpanzees, might have some understanding of other individuals' visual perspectives. Burkart and Heschl then designed another experiment to probe the mechanisms underlying the marmosets' behavior. They trained the marmosets to approach one of two pieces of food that had been placed in front of a human. Half of the marmosets were trained to approach the food that the human was looking at; the other half were trained to approach the food that the human was *not* looking at. The latter group learned this task much more quickly than the former, suggesting that marmosets are naturally more inclined to take food that is being ignored than food that is probably being defended. It therefore seems likely that, in the first experiment, the marmosets had approached the food visible only to themselves not because they recognized what their rival could or could not see, but because they were avoiding the food at which their rival was looking. Rhesus macaques might have stolen the grape from the unseeing person for a similar reason.

To summarize, there is no strong, conclusive evidence that monkeys, apes, or other animals recognize other individuals' knowledge. Even if we grant that they do, this recognition—as it is in young children— seems tacit and inaccessible to explicit manipulation in a variety of different contexts (Onishi and Baillargeon 2005). Many animals, however, appear to recognize other individuals' intentions and motivations. As we discuss below, this recognition is not just revealed in competitive contexts, but also extends to cooperative ones.

Inferring the intentions and dispositions of others: evidence from vocalizations

Clearly, many of the experiments we have just described cannot easily be transferred to the wild. Among many other constraints, we would guarantee the complete demolition of our camp kitchen if the baboons ever learned to associate us with food. We have therefore had to address the question of intent more indirectly, through their vocalizations.

During conversation, humans routinely make inferences about the motives and beliefs of their intended recipients (Chapter 10). When someone else is speaking, we also make inferences about who the intended recipient is. Courtrooms, for example, would be chaotic and confused indeed if every member of the jury shouted out his own reply whenever the prosecutor asked the defendant a question. Baboons, too, act as if they recognize the intended recipient of someone else's calls. Baboon groups are noisy, tumultuous societies. At any given moment, several females may be grunting as they interact with other females' infants, a male may be grunting as he approaches a more dominant male, another female may be giving threat-grunts at a screaming juvenile, and one of her sisters may be giving threat-grunts in vocal support. If a baboon interpreted every vocalization she heard as directed at herself, she would soon be a nervous wreck, unable to feed or sit for more than seconds at a time. Clearly, she must be able to infer when a vocalization concerns her (or one of her relatives), and when it does not. Even an egocentric baboon has to know that it is not always about her.

Inferences about the "directedness" of vocalizations are probably often mediated by gaze direction and relatively simple contingencies. In a typical interaction, a dominant female approaches two lower-ranking females, one of whom has an infant, and utters a grunt. The mother with the infant remains seated, while the other female moves away. How do they decide whether to stay or leave? Typically, the grunting female is looking at the mother, and both listeners know that grunts are correlated with infant handling (Chapter 5). As a result, both females infer that the grunt is directed at the mother, and she stays where she is while the other moves away.

When accompanying visual signals are absent, however, the interpretation of vocal signals may be more difficult. Baboons often forage in woodland where other individuals can be heard but not seen and where animals approach each other from a variety of directions and angles. Because grunts are individually distinctive, listeners can readily identify an unseen caller. Unless she can see the caller, though, the listener may find it difficult to determine whether she or another baboon is the intended recipient. To examine how baboons make this decision, we return to our study of "reconciliatory" grunts.

Inferences about dominants' vocalizations

As we discussed in Chapter 5, baboon grunts function to facilitate and mediate social interactions and to reconcile opponents after aggression. Grunts both reduce the anxiety of subordinate females and influence their subsequent behavior. But how do females recognize the reconciliatory function of grunts? The rich interpretation assumes that females attribute emotions like anxiety to others. The dominant female recognizes that the fight has upset the subordinate female, so she grunts in order to alleviate her anxiety—even though, being dominant, she feels no anxiety herself.

The simpler explanation makes no recourse to a theory of mind. It argues only that dominant females grunt to subordinates in order to influence their behavior. Through experience, dominant females learn that subordinate females are less likely to move away from them when they grunt than when they remain silent, so they grunt. Subordinate females, being exquisitely sensitive to contingencies, learn that grunts are associated with friendly behavior, so they do not move away when their former opponent grunts to them.

Our playback experiments demonstrated that subordinate females were more likely to approach their former opponent and tolerate her approaches if they heard her grunt than if they did not. But when they heard a different dominant female's grunt, they did not approach that female. They behaved as if they assumed that this grunt was directed at someone else. These results were consistent with the hypothesis that baboons have a rudimentary understanding of other individuals' intentions toward themselves, but simpler explanations could still not be completely ruled out. In particular, it remained possible that a recent interaction with a specific individual—*any* sort of interaction, even a fight—might simply prime females to react to that individual's vocalizations.

Anne Engh and we therefore designed a playback experiment to test whether a listener's responses to another female's vocalizations are influenced by the nature of their recent interactions. The experiment again followed a within-subject design. In the first condition, a subordinate female was played the threat-grunts of a dominant female shortly after that female had threatened her. Because females threaten their former victims again within five minutes following approximately 14% of all fights, we expected that listeners would interpret these threat-grunts as an indicator of renewed aggression directed *at them*. In the second condition, the same subordinate female was played the same dominant female's threat-grunts shortly after the two had groomed. Because females almost never threaten a recent grooming partner, we expected that in this case listeners would interpret the call as directed at someone else.

If baboons' responses to threat-grunts are simply the result of priming because of a recent interaction, subjects' responses after being threatened should have been the same as after being groomed. If, however, listeners take into account the nature of recent interactions when making inferences about the intended recipient of a call, they should have interpreted the two threat-grunts differently—directed at them after aggression, and directed at someone else after grooming. Specifically, we predicted that subjects would respond more strongly to threat-grunts after receiving aggression than after a grooming bout. We also predicted that a subject would be less likely to approach her former opponent and more likely to retreat from her approaches after being threatened by her than after being threatened by a different dominant female. In contrast, when the subject heard the same female's threat-grunts after grooming with her, she should have been just as likely to approach and to tolerate her approaches as she was after having been groomed by a different dominant female.

Indeed, subjects did respond more strongly to a dominant female's threat-grunts after being threatened by her than after grooming with her. After aggression, subjects were quicker both to look toward the speaker and to move away from the area. In the ensuing 15 minutes, they were also less likely to come near their former opponent than they were after they had been threatened by a different female. In contrast, when subjects heard the same female's threat-grunt after a grooming bout, they were just as likely to tolerate her approaches as they were after a grooming bout with a different female. Finally, subjects were significantly more likely to approach and to tolerate the approaches of the dominant female if they heard her threat-grunts after grooming than after a threat (Engh et al. 2006c).

As in the experiments on reconciliatory grunts (Cheney and Seyfarth 1997), subjects' responses were specific to their former opponent.

Hearing their aggressor's threat-grunt did not affect the likelihood that subordinate subjects would approach another, uninvolved dominant female or the likelihood that they would be supplanted when approached. Taken together, therefore, these two experiments suggest that female baboons make inferences about the intended target of a vocalization even in the absence of visual cues, and that the nature of prior interactions affects subsequent behavior. After a fight, the subordinate assumes that the dominant has aggressive intentions toward her. After grooming, or after hearing a reconciliatory grunt, she draws the opposite conclusion.

It seems likely that baboons make inferences about the intended target of a call whenever they hear any vocalization. Recall the playback experiments conducted by Wittig and Crockford on vocal alliances (Chapter 6; Wittig et al. 2007b). When subjects were played the threat-grunts of their aggressor's relative soon after being threatened, they avoided members of their aggressor's matriline. In contrast, when they heard the same threat-grunts in the absence of aggression, they ignored the call and acted as if they assumed that the call was directed at someone else.

Similarly, when subjects heard the "reconciliatory" grunt of their aggressor's relative after a fight, they were more likely to approach both their aggressor and the relative whose grunt they had heard. They did not venture to do so, however, if they had heard the "reconciliatory" grunt of another, unrelated female. Here again, subjects behaved as if they believed that a grunt from their aggressor's relative must be directed *at them,* as a consequence of the fight. An unrelated female's grunt was deemed irrelevant. What is especially interesting in these experiments is that subjects inferred that they were the target of the vocalization even though they had not recently interacted with the signaler, but with her relative. They could only have done so if they recognized the close bond that existed between the two females.

There is an intriguing parallel between these results and recent neurophysiological research. In primates, faces and voices are the primary means of transmitting social signals, and monkeys recognize the correspondence between facial and vocal expressions (Ghazanfar and Logothetis 2003). Presumably, visual and auditory signals are somehow combined to form a unified, multimodal percept in the mind of a monkey. In a study using positron emission tomography (PET), Gil da Costa and colleagues (2004) showed that when rhesus macaques hear one of their own species' vocalizations, they exhibit neural activity not only in areas associated with auditory processing but also in higher-order visual areas, including STS. Auditory and visual areas also exhibit significant anatomical connections (Poremba et al. 2003).

Ghazanfar and colleagues (2005) explored the neural basis of sensory integration using the coos and grunts of rhesus macaques as stimuli. They found clear evidence that cells in certain areas of the auditory cortex are more responsive to bimodal (visual and auditory) presentation of species-specific calls than to unimodal presentation. Although significant integration of visual and auditory information occurred in trials with both vocalizations, the effect of cross-modal presentation was greater with grunts than with coos. The authors speculate that this may occur because grunts are usually directed toward a specific individual in dyadic interactions, whereas coos tend to be broadcast generally to the group at large. The greater cross-modal integration in the processing of grunts may therefore have arisen because, in contrast to listeners who hear a coo, listeners who hear a grunt must immediately determine whether or not the call is directed at them.

The ability to distinguish signals directed at oneself from those directed at someone else appears to be widespread in animals. Studies of "eavesdropping" in birds indicate that listeners readily distinguish between songs directed at a third party as opposed to ones directed at them (Naguib et al. 1999; Peake et al. 2001, 2002; Chapter 7 of this volume). To date, however, most of the evidence for this ability has come from studies in which individuals are interacting with only one or a few other animals, and when factors such as the location of the signaler and the nature and pattern of his song provide information about the intended recipient. The challenge of inferring both the intended target of a signal and the signaler's probable behavior may be considerably more difficult in large social groups.

It is unlikely that baboons use simple distance- or sight-based rules of thumb to determine the intended target of a call. Because baboons are often sitting in close proximity to others, a female cannot simply assume that all calls given by nearby signalers are directed at her, nor can she assume that calls given by distant signalers are not. Indeed, as the experiments with contact barks showed, baboons often respond vocally to signalers who are out of sight and widely separated from them.

In sum, when deciding "who, me?" upon hearing a vocalization, baboons must take into account the identity of the signaler (who is it?), the type of call given (friendly or aggressive?), the nature of their prior interactions with the signaler (were they aggressive, friendly, or neutral?), and the correlation between past interactions and future ones (does a recent grooming interaction lower or increase the likelihood of aggression?). Learned contingencies doubtless play a role in these assessments. But because listeners' responses depend on simultaneous consideration of all of these factors, this learning is likely to be both complex and subtle.

Furthermore, explanations based solely on learned contingencies cannot easily explain some aspects of baboons' behavior. For example, in the experiments that mimicked vocal reconciliation, subjects who heard their opponent's grunt following a fight were even more likely to approach their opponent than they were under baseline conditions, in the absence of a fight. If listeners' responses were guided solely by learned contingencies, they should have associated the call only with a low probability of aggression. Hearing the call should have returned their behavior to baseline tolerance levels, but it should not have induced them to approach their former opponent. Instead, females acted as if they interpreted their opponent's grunt as targeted specifically at *them*, as a directed signal of benign intent. They therefore deliberately sought out their opponent.

Inferences about subordinates' vocalizations

The experiments just described may also help us to understand the function of some other puzzling vocal exchanges. Thus far, we have concentrated on baboons' inferences about dominant females' vocalizations. But subordinate animals also use vocalizations to manage their social interactions. Often, these grunts seem to function as a kind of announcement that the female is doing or is about to do something that she recognizes is above her station and might otherwise not be tolerated. For instance, a low-ranking female who wishes to inspect or handle the infant of a higher-ranking female typically approaches the female slowly and judiciously, grunting all the while. She then brings her eyes to the level of the infant, gazes into its eyes, and grunts repeatedly before reaching out a tentative hand. The worse the dominant female's reputation for unprovoked malice, the more prolonged the supplication process is.

Traditionally, the subordinate female's behavior would be explained in terms of a learned contingency: the female has learned that she is more likely to be allowed to touch the baby if she grunts. Our experiments suggest an alternative explanation: the subordinate female grunts to signal her benign intent, and the prolonged supplication process allows the female to assess the dominant female's disposition toward her. Learned contingencies still play a role, but so do inferences about intent.

In a similar manner, low-ranking females often seem to be attempting to assess higher-ranking individuals' dispositions by "announcing" their presence to them. Here is an example. Balo, the 14th-ranking female in the group at the time, was feeding alone in a jackalberry tree when Comet, the seventh-ranking female, approached. Balo grunted. Comet, who had not yet noticed Balo, stopped walking and looked up into the tree. She grunted and lipsmacked to Balo, who grunted and lipsmacked in return. Comet then entered the tree and fed near Balo. Why did Balo reveal herself when Comet had not yet detected her, especially when Balo risked eviction from the tree by doing so? Our observations indicate, somewhat paradoxically, that low-ranking individuals who fail to announce their presence to higher-ranking ones are more likely to be supplanted from a desirable resource than those who announce themselves.

Similar observations have been made in other monkey species. In an experiment conducted with rhesus macaques on Cayo Santiago, for example, Hauser (1992b) found that low-ranking females who discovered food were less likely to be threatened by higher-ranking individuals if they produced a "food" call than if they remained silent. This harassment was initially interpreted as evidence that monkeys can recognize when others are trying to deceive them, but this explanation seems unlikely given monkeys' and apes' inability to attribute beliefs to others and the apparent lack of punishment in other primate species (e.g., Clark and Wrangham 1994). A more plausible interpretation is that food calls function to announce possession. Capuchin monkeys who produce food calls are also less likely to be challenged by more dominant individuals than monkeys who remain silent (Gros-Louis 2004). Experiments suggest that this occurs not because dominants are punishing subordinates but because the calls serve to warn listeners of the signaler's willingness to defend her food. As a result, unless they are strongly motivated to take the food, listeners refrain from harassing her. Similar "respect for ownership" has been observed in other monkeys and other contexts (Kummer et al. 1974).

If this interpretation is correct, signals of "announcement" by subordinate animals serve both to broadcast intent and to assess the intentions of others. In some contexts, they act as mildly aggressive signals that denote possession; in others, they act to open the process of negotiation. On many occasions, for example, we have observed a juvenile or low-ranking female almost bump into a higher-ranking animal as she moves through tall grass. As the lower-ranking female screeches to a halt, she often grunts to the dominant. If the dominant grunts in return, the subordinate will resume walking and pass close by the dominant (Fig. 36). But if the dominant remains silent, the subordinate will detour or even turn around. Again, it is as if the two are assessing each other's motives and moods.

Figure 36. Unless Selo grunts to a female when they come into proximity of each other, the other female will almost invariably move away. Photograph by Anne Engh.

To summarize, although baboons and other monkeys probably do not recognize when someone is attempting to manipulate their beliefs, they may recognize when someone is attempting to manipulate their intent. They integrate social cues, gaze direction, and call type when making these assessments and when announcing their intentions to others. A rudimentary understanding of intentions and motives represents a crucial first step toward a communication system like language, in which speakers and listeners routinely assess each other's motives, beliefs, and knowledge.

Behaviorists will object even to this watered-down interpretation of mental state attribution in animals, and rightly so. None of the experiments we have described allows us distinguish between the implicit understanding of an intention as a mental state and the implicit understanding of an intention to behave in a certain way. When her recent opponent grunts to her, does a baboon attribute a friendly mental state to her opponent, or does she only recognize that her opponent intends to be friendly? The distinction is so subtle that it may be impossible to discern through behavior alone and may in any case be functionally meaningless. We know that baboons and other animals are astute observers of other individuals' behavior and that they are very sensitive to subtle contingencies (Povinelli and Vonk 2004). Although the neuroanatomical evidence predicts that they should also be sensitive to intentions, goals, and motives, no experiment has yet proven so definitively. Baboons and other animals excel at reading and predicting behavior. The jury is still out on whether they also excel at reading and predicting simple mental states.

Recognizing animals' intentions and dispositions toward other individuals

By now it should be clear that baboons and other monkeys have quite sophisticated knowledge about other individuals' social bonds. This knowledge may also enable them to make quite sophisticated assessments about other individuals' intentions. Consider the following anecdote.

Leko had given birth two days earlier, during a period when three infants had recently been killed by an infanticidal male. The group was being pursued by a larger group—a context in which infanticide often occurs—and the males were displaying and chasing females. We suddenly saw Leko being chased by an adult male. The male chased Leko

Figure 37. A male forces a female to walk the plank. Photograph by Dawn Kitchen.

for over a hundred meters until he finally tackled her to the ground and bit her (Fig. 37). We were certain that we were witnessing an infanticide attempt, but oddly none of the other baboons screamed or barked, as they normally do during such incidents. We then recognized the male as Loki, the male with whom Leko had a close friendship and with whom she had consorted when she became pregnant. Apparently, the baboons had also recognized that it was Loki chasing Leko and that Loki was unlikely to harm Leko's infant, even though all of his behavior was consistent with an infanticide attempt.

Of course, an anecdote like this does not allow us to conclude that the baboons recognized Loki's intentions. We can only say that baboons' recognition of other individuals' social relationships enables them to predict with great accuracy what other individuals are likely to do, even when their ostensible behavior suggests otherwise. It is hard to imagine how a baboon could make this assessment without at least some tacit inferences about others' motives.

Somewhat to our dismay, the baboons also seem to understand *our* dispositions, and go to some lengths to irritate us. In 2005, we were visited by two photographers from the BBC, one of whom carried a large tripod to steady his camera. On their third day out with the baboons, Drongo, an insouciant adolescent male, sauntered over to the cameraman and looked him straight into the eye for several seconds. He then strode toward the tripod, which was standing about 10 meters away, sat down, looked at the cameraman again, and pulled roughly on one of the cords dangling from the camera. Drongo did not approach or look at any of the other four people standing nearby. He evidently associated the tripod with a specific human and seemed to recognize that his behavior would provoke a strong response if he first announced his intentions to the appropriate person.

But although baboons seem adept at reading other individuals' immediate motives and intentions, they often seem surprisingly naïve when it comes to monitoring others' intentions over the long term. For example, when a high-ranking male immigrates into the group, lactating females immediately seem to recognize him as an infanticidal threat (Chapter 4). Their stress levels increase and they run screaming from him if he approaches. But as time passes, the females make fewer efforts to avoid the new male. If he commits infanticide, they flee from him for the next week or two. However, if no infants have recently been attacked, some females eventually begin to feed and even rest near the immigrant, especially if their male friend is nearby. Even though the immigrant remains a threat to lactating females for over a year, their efforts to avoid him diminish with time. Similarly, although all group members respond with loud screams and barks when a male attempts to kill an infant, they all resume feeding with no apparent lingering concern soon after the incident ends.

Why do lactating females not make more of a concerted effort to avoid a potentially infanticidal male? And why does an infanticide attempt produce—at least ostensibly—only temporary outrage? One explanation is that a baboon's ability to read another's intentions depends to some degree on memory of his recent behavior. As a result, a potentially infanticidal male who is feeding peacefully and has not made any attacks on infants for several weeks is not regarded as someone with long-term nefarious motives. On the other hand, lactating females may very well understand that the male remains dangerous, but they simply cannot afford to spend the next 12 months running screaming from him whenever he approaches. In the absence of an interview, it is impossible to tell.

Imitation and teaching

Imitation

Almost all animals learn novel tasks more easily if they can observe a knowledgeable demonstrator. Langen (1996) trained individual magpie jays (*Calocitta formosa*) to pry open a door on a box that contained food. Subsequently, birds whose social group included a "demonstrator" learned how to open doors much more rapidly than birds whose groups did not. Indeed, birds in groups that lacked a demonstrator did not even realize that there was food in the boxes. Similarly, in captivity many monkeys can learn to use rudimentary tools to obtain food, and they do so more quickly and accurately in the presence of a demonstrator (e.g., Fragaszy and Visalberghi 1989; Bugnyar and Huber 1997; Caldwell and Whiten 2004; reviewed by Tomasello and Call 1997). If monkeys learned like humans do, it would be safe to assume that the monkeys learn to perform the tasks by watching the demonstrator and imitating his actions. This would imply that the monkeys understand the demonstrator's intentions and goals. But this does not seem to be the case.

Most animal species show very little evidence of purposeful copying by imitation. In the laboratory, monkeys are attracted to tools and often begin fiddling with them after observing another monkey do so, suggesting that social companions enhance and facilitate tool use. But learning about a tool's use through "social facilitation" typically requires extensive practice through trial and error. As a result, different individuals adopt different idiosyncratic styles, and the spread of the skill is very slow. Even in the famous case of potato washing by Japanese macaques, only 11 of 25 monkeys adopted the practice over a three-year period (Nishida 1987). Thus, trial and error learning, facilitated by proximity to companions who had already acquired the skill, appears to have driven the behavior's spread.

Although many monkey species can learn to use tools in captivity, there are very few examples of tool use in the wild. The New World capuchin monkeys are the only monkeys that regularly use sticks or stones to pry into trees or break open nuts under natural conditions (Fragaszy et al. 2004). Capuchins also have comparatively large brains compared to other monkeys (Rilling and Insel 1999). The relative lack of spontaneous tool use in most monkey species suggests that monkeys have difficulty representing the task at hand and recognizing the relation between actions and objects (Visalberghi and Fragaszy 2002).

In the wild, baboons seem to use tools only in aggressive contexts. When displaying, male baboons occasionally wave or throw sticks in the direction of their rivals. Whether they recognize the potential function of these weapons, though, seems doubtful. When one group of baboons in Namibia dislodged stones from a cliff when they were disturbed by humans, they did so not only when the people were under the cliff but also when they were too far away to be struck (Hamilton et al. 1975). At Gombe, where chimpanzees compete with baboons for food, chimpanzees throw branches at baboons (Goodall 1968). Baboons, however, never throw objects at chimpanzees.

Chimpanzees and orangutans are different. In captivity, these apes attend closely to a demonstrator when learning to use tools to open boxes, and they require very few trials to learn to copy his actions. They seem to recognize the intentions and goals of the demonstrator, and they rapidly learn a tool's function from attending to his behavior (reviewed by Tomasello and Call 1997). Although they do not copy the demonstrator's exact motor patterns as slavishly as children do, they do tend to conform to his technique (Whiten 2002; Whiten et al. 2005).

Under natural conditions, chimpanzees and orangutans also use a variety of tools for different purposes (reviewed by Tomasello and Call 1997; van Schaik et al. 2003; van Schaik 2004). We will not attempt to review the literature on ape tool use here, except to say that different populations of chimpanzees and orangutans use different kinds of tools for different purposes, and that the use of specific tool types appears to be socially transmitted. For the purposes of our discussion, two points are relevant. First, in marked contrast to monkeys, no population of chimpanzees has been reported *not* to use tools. Second, unlike monkeys, chimpanzees and orangutans often show foresight and planning in selecting and modifying tools in advance of their use. Before fishing for termites, chimpanzees often search some distance from the termite mound to find an appropriate prodding stick and strip the bark from it (Goodall 1968). Similarly, when preparing to crack open nuts, chimpanzees must carry both stones and nuts to suitable anvils. Often, this means that a chimpanzee will carry both nuts and stones over considerable distances before beginning a nut-cracking session (Boesch and Boesch 1984).

In their ability to plan, understand a tool's function, and appreciate a demonstrator's goals, then, apes are strikingly different from most monkeys. This is not to say, however, that tool use and manufacture are unique to apes, or that monkeys are completely incapable of imitation.

First, some birds seem to be almost as adept at tool use and planning as chimpanzees. Under natural conditions, New Caledonian crows (*Corvus moneduloides*) manufacture and use tools to pry insects from crevices (Hunt and Gray 2004). They often fly a considerable distance from their feeding tree to find a suitable prodding stick. In many instances they prepare the stick much as chimpanzees do, by breaking off the tip and stripping the leaves and bark. Betty, a particularly resourceful crow and a late resident of Oxford, spontaneously learned how to bend a piece of wire into the shape of a hook, which she then used to pull up a bucket containing food from a well (Weir et al. 2002).

Second, a recent experiment investigating "cognitive" imitation in monkeys challenges the view that apes are the only primate species that can imitate others (Subiaul et al. 2004). In this experiment, two rhesus macaques had to learn to respond in correct order to a series of four pictures displayed in random positions on a computer screen. A typical series of pictures might occur in the order: (1) *bird;* (2) *car;* (3) *house;* (4) *dog*. The monkeys' job was to press each picture in the correct order. After completing the sequence correctly, the monkeys received a reward. Both monkeys soon became proficient at this task, and when they were presented with a news series of pictures they learned its correct order through trial and error.

Next, the experimenters presented each monkey with 50 new lists and assigned each list to one of four different conditions. In one condition, the subject was alone and had (as before) to work out the list's correct order by himself. In a second condition, a second monkey sat in the test chamber adjacent to the subject and worked on a different list depicting different pictures. In a third condition, a computer in the adjacent chamber worked on the same list as the subject's and repeatedly demonstrated the list's correct order on its screen. Finally, in the fourth condition, a "demonstrator" monkey who had already learned the list's correct order sat in the adjacent chamber and worked on the same list as the subject. In other words, in the first two conditions the subject could learn the list's correct order only through trial and error. In the latter two conditions, though, the subject could also learn the list by copying—either the computer, in the third condition, or the demonstrator monkey, in the fourth.

Both monkeys learned the list significantly faster in the presence of the demonstrator than in the other three conditions, indicating that they were imitating the demonstrator. Simple social facilitation could be ruled out, because if the monkeys had only been attracted to the task because the adjacent monkey was also working on a list, they should have learned the list just as fast when the adjacent monkey was working on a different list (the second condition). Instead, the monkeys seemed to understand that, to complete the list correctly, they had to follow the lead of the monkey who was working on the same list as they were, and respond to the items in the same order as he did. Furthermore, they had to copy not just his first choice, but also his subsequent ones.

One reason why this experiment may have succeeded where so many other tests of imitation in monkeys have failed is that it did not require the monkeys to imitate a motor pattern or to learn the purpose of the task. The monkeys already knew what they needed to do—they simply did not know the correct list order. The task required the monkeys to focus not on the problem as a whole—only on the demonstrator's specific choices. Their ability to do so is perhaps not unexpected, given neural evidence that monkeys are highly sensitive to goal-directed actions and gazes (see above).

Teaching

Even more than imitation, teaching requires the ability to attribute a mental state different from one's own to others, because the instructor must understand how and in what respects his knowledge diverges from his pupil's. Although human cultures vary considerably in the emphasis put on teaching, some degree of teaching occurs in all human societies. Evidence for teaching by nonhuman primates, however, can be summarized by one word: scant. The anecdotes from captivity are both provocative and difficult to interpret. For example, Kohler (1925) reports an incident in which Sultan, a male chimpanzee adept in the art of stacking boxes to obtain bananas, watched with increasing frustration as more hapless group members failed to solve the problem. Finally, he ran into the room, stacked the boxes, and ran out of the room without attempting to obtain the bananas himself. Kohler concluded that Sultan understood that the other chimpanzees could not solve the problem and was motivated to demonstrate the solution to his unenlightened companions.

Given such intriguing anecdotes, it is sobering to discover that the most common approximations to teaching in free-ranging monkeys and apes occur in the form of punishment for some social transgression. Mothers aggressively interfere in rough play between their offspring and other juveniles, push infants from their nipples during the weaning period, and retrieve their infants from females who are handling them roughly. But these corrective actions seem to derive less from pedagogical intent than from an attempt to remedy a situation that is unpleasant to the actor.

Outside this punitive domain, examples of teaching are even rarer. Monkeys and apes are surrounded by dangerous predators and poisonous foods, and infants and juveniles soon learn which animals and foods to avoid. They do so, however, without benefit of teaching. When infant vervet monkeys begin to give alarm calls, they often make "mistakes" and give alarm calls to species that pose no threat to them (Chapter 10). Adults sometimes respond to these mistakes, albeit briefly. If, for example, an infant gives an eagle alarm call in response to a small hawk, adults will glance up and then go back to doing whatever they had been doing. But if an infant gives an eagle alarm call in response to a martial eagle (*Polemaetus bellicosus*), a true predator, adults will look up and give alarm calls themselves. At first glance, these corroborating alarm calls seem to be deliberately instructive, because they reinforce the infant's correct response. Adults, though, are just as likely to give corroborating alarm calls after another adult has given a correct alarm call as they are after an infant has. Even though infants make many more errors than adults, adults make no special effort to reward them if they are correct. The infant is left to infer for himself that the correct alarm calls are the ones matched by adults (Cheney and Seyfarth 1990).

The relative absence of teaching in nonhuman primates is particularly striking in the case of chimpanzee tool use. It takes many years for juvenile chimpanzees to learn how to fish for termites or crack open palm nuts with a rock. Mothers often aid their offspring in this enterprise by tolerating them at the feeding site, allowing them to grab and handle their tools, and sometimes also giving their tools to them. Examples of explicit teaching, however, are rare; learning is passive and involves little active intervention (Boesch 1991; Lonsdorf 2006).

Empathy

Both human and nonhuman species show physiological responses to the loss of close companions, which we label as the emotion of grief. In humans, bereavement and feelings of loneliness are associated with increased cortisol production, declines in immune responses, and, in some cases, increased mortality (Chapter 5). In a variety of animals ranging from rodents to primates, social isolation and separation from a close companion produce an elevated stress response. Baboons experience the same kinds of physiological responses when they lose a preferred companion, even when they have other kin or close companions with whom they can still interact.

Grief is an egocentric emotion, like fear; it does not require any ability to attribute mental states to others (Fig. 38). It is certainly possible to feel grief or a sense of loss without recognizing that others might feel the same way. In contrast, empathy requires that an individual be able to recognize emotions like grief or fear in others even when she is not experiencing these emotions herself. It demands that she deliberately

Figure 38. Baboons seem not to recognize when others are experiencing grief. Photograph by Keena Seyfarth.

imagine herself in another individual's position while still dissociating her own mental states from her companion's. In humans, representations of emotions like pain, disgust, and shame in others activate many of the same areas of the brain as those activated when we experience or imagine ourselves experiencing the same emotions. Other areas of the brain, including in particular the right inferior parietal cortex and the prefrontal cortex, allow us to detach our own emotions and knowledge from others' (Decety and Jackson 2004, 2006). Thus, although we use much of the same neural architecture to understand our own and others' mental states, we are nonetheless able to maintain a degree of separation between them.

Chimpanzees

Chimpanzees are often described as showing compassion for others. In the wild, they have been reported to build nests and bring food to an injured relative (Goodall 1986). Chimpanzee mothers have also been observed attending to the wounded or paralyzed limbs of their offspring (Goodall 1986). Clearly, they seem to show compassion toward incapacitated and dependent companions. But do chimpanzees ever show empathy? There is considerable disagreement on this issue. Most examples are anecdotal (e.g., Preston and de Waal 2002) and subject to a wide variety of interpretations. And for each anecdote that seems to demonstrate empathy there are several counterexamples. In fact, two experimental studies designed explicitly to test whether chimpanzees show concern and regard for others concluded that they do not (Silk et al. 2005; Jensen et al. 2006).

In these experiments, conducted on captive group-living chimpanzees at several different sites, individuals had the opportunity to share food with another group member at no cost or inconvenience to themselves. The actor was seated in a cage next to or across from a second cage containing a familiar group companion. The actor had the choice of selecting one of two handles that delivered food. If he pulled one handle, a tray of food was delivered only to himself. If he pulled the other handle, identical food rewards were delivered simultaneously to both him and his companion. The actor's choice did not affect his own payoff, because he received the same amount of food in either case. However, even though some subjects alternated as both actors and recipients, and even though all had lived with their companion for many years, they were no more likely to choose the second option than the first. Although they were highly motivated to obtain food for themselves, they were indifferent to the welfare of their companion. They were not spiteful—they did not consistently withhold food from their companion. Instead, they simply failed to take the other individual's perspective into account. In contrast, children as young as 18 months not only recognize when another person needs help but also go out of their way to help the person achieve his goal (Warneken and Tomasello 2006).

This difference between chimpanzees and children is striking. It seems unlikely that the disparity is due entirely to a failure on the part of chimpanzees to recognize other individuals' intentions, emotions, and motivations because, as we have seen, several experiments have now suggested that chimpanzees and even monkeys have some ability to do so. Instead, chimpanzees and other nonhuman primates may lack a capacity abundantly evident in even very young children—the motivation to collaborate and to share goals, emotions, and knowledge with others (Tomasello et al. 2005; Chapter 12 in this volume). We do not mean to suggest that chimpanzees do not cooperate with one another or that they never engage in collaborative activities—they do. Male chimpanzees jointly patrol and defend their community's territory, and they also participate in cooperative hunts (Goodall 1986; Boesch 2002). But when chimpanzees hunt, they do not consistently coordinate their respective roles, nor do they punish those who fail to share the catch. They appear to be much less sensitive than humans to the psychological mechanisms underlying cooperation.

Monkeys: Anecdotal evidence

The search for humanlike empathy in baboons and other monkeys is less controversial than it is in apes, perhaps because there is very little evidence for empathy in these species. Despite their close social bonds, monkeys do not share food with sick or injured companions or attend to the old and disabled. Even when mothers carry sick or dying infants, they do not treat them very differently from the way they treat healthy infants. Although monkeys do sometimes groom and examine others' wounds, they appear to treat injuries as anomalies or objects of interest rather than as handicaps that require adjustments in their own behavior. And, as we have mentioned, baboon mothers often show a surprising lack of concern for their offspring's anxiety and distress during water crossings or at other times of separation. Similar observations have been in made in other monkey species. For example, Japanese macaques scratch themselves at high rates after being attacked—a behavior that reflects their anxiety. But they do not scratch themselves more after their infants are attacked (Schino et al. 2004).

Our ability to measure glucocorticoid levels in baboons presented us with an ideal opportunity to assess their capacity for empathy. In addition to experiencing increased stress at the loss of a close companion, humans often show an increase in glucocorticoid levels in response to grief or anxiety in others. Health care workers who attend to sick or traumatized patients often experience "compassion fatigue" and elevated stress profiles (Figley 1995). Baboons have many opportunities to experience vicarious stress. There is little evidence, however, that they do so, even when their own fitness is affected.

When a potentially infanticidal male immigrates into the group, glucocorticoid levels in cycling females are unaffected. Even when the male begins to commit infanticide, the stress levels of cycling females do not rise significantly. This is perhaps not surprising, because the offspring of cycling females are not in danger. In many cases, though, these cycling females have daughters, mothers, or sisters who are lactating at the time and whose infants are at considerable risk. Because these infants are close relatives, their deaths would reduce the cycling females' own overall fitness. Nevertheless, the cycling females' stress levels are unaffected. Although it could be argued that it is maladaptive for female baboons to respond to every indirect threat to their fitness, the lack of empathy is puzzling. It is as if baboons do not respond to a threat unless it has the potential to have an immediate impact on them. The indirect impact is not recognized.

Selo, for example, was cycling when Spock immigrated into the group and assumed the alpha position. Selo's daughter, Palm, had a young infant at the time and assiduously avoided Spock. Selo soon formed a sexual consortship with Spock. Although Spock followed her tenaciously, Selo continued to seek out Palm as a grooming companion. As a result, Palm was often in much closer proximity to Spock than she preferred to be. Within two months of his arrival in the group, Spock killed Palm's infant. We watched Selo's behavior with considerable exasperation. How could she be so insensitive to Palm's predicament?

The adults appeared to respond in a similarly callous manner during the Lord of the Flies incident. Despite being able to hear their offspring's agitation and distress, there was no evidence that the adults empathized with the youngsters' dilemma. None of the females showed an elevated stress response during the week of the separation. They knew where their offspring were, they knew that they could swim relatively easily to the island where their offspring were stranded, so what was all the fuss about?

This is not to say that baboons are entirely unresponsive to a companion's injury or loss. When her infant dies, a female baboon will often continue to carry the body for as many as 10 more days, cleaning the corpse of maggots and brushing flies away from it. As the corpse decays and mummifies, she begins to leave the body for increasing lengths of time before finally abandoning it. It is as if the mother continues to respond to the corpse as her infant even after it has lost all resemblance to a baboon.

In the minds of other group members, the infant's status seems to change soon after it dies: they cease to treat it as an infant. They inspect the corpse with great curiosity, but they seldom attempt to handle it. When they approach it they rarely grunt, as they would if the infant were still alive. Nevertheless, they appear to recognize that the corpse still belongs to the mother. They approach the mother cautiously and do not attempt to try to take the corpse from her. When the mother moves away from the body, other group members grunt to her, and a close relative or male friend often guards the body until she returns. Even after the mother finally abandons the blackened, mummified corpse, the baboons continue to threaten any human who attempts to approach it in the vain hope of obtaining a tissue sample for DNA analysis.

What goes on in a baboon's mind as she carries her dead infant? What goes on in other group members' minds? We will not attempt to suggest that baboons have a concept of death, or that they ruminate about the meaning of life. Nevertheless, baboons do seem to recognize that a corpse is something of a baboon manqué. Although not treated as a living baboon, it still seems to be regarded as something that belongs to a particular individual and family, and group members cooperate to defend the corpse. But much as we might be tempted to interpret the baboons' behavior as empathy, it seems more likely that it simply reflects their "respect for ownership"—a reluctance to challenge an individual, or kin group, whose motivation to defend a possession is high.

Monkeys: Experimental evidence

Even when monkeys appear to go out of their way to alleviate a companion's distress, they seem to do so not because they feel empathy independent of their own concerns but because they associate their companion's distress with their own feelings of distress. In an early experiment specifically designed to examine whether monkeys would respond to another monkey's distress, macaques were trained to pull one chain in response to a red light and another in response to a blue light, to obtain a food reward. After the monkeys had learned to pull the appropriate chain when presented with the appropriate signal, the apparatus was rigged so that a monkey in an adjacent cage received an electric shock each time one of the chains (say, the chain associated with the red light) was pulled. Most of the monkeys soon stopped pulling the chain that delivered the shock, even though they deprived themselves of a food reward by so doing. They were especially likely to avoid the chain if they had, in the past, received shocks themselves (Masserman et al. 1964; Wechkin et al. 1964).

Although the monkeys' responses in these experiments might at first be seen as evidence for empathy, it seems more likely that they became distressed when they saw the other monkey being shocked because it evoked memories of being shocked themselves. They avoided the chain because it was associated with a negative experience. Their apparent concern for the other monkey's welfare was therefore inextricably linked to concern for themselves (Silk in press b).

How, then, should we interpret kin-mediated reconciliation in baboons? In Chapters 5 and 6, we described experiments showing that female baboons treat the "reconciliatory" grunt of their opponent's kin as a proxy for reconciliation by their opponent (Wittig et al. 2007a). In chimpanzees, such third-party reconciliation, or consolation, has been interpreted as evidence for empathy (Preston and de Waal 2002; Aureli and de Waal 2000). The apparent absence of consolation in monkeys has, in turn, been taken as evidence that chimpanzees, but not monkeys, are able to use representations about their own mental states to understand the mental states of others. As we have seen, however, baboons do "console" their relatives' victims, and at rates similar to those observed in chimpanzees.

Although it is certainly possible that kin-mediated reconciliation in baboons might involve the ability to empathize, simpler explanations seem (once again) more likely. As we have discussed, group life is essential for baboons and other monkeys. It therefore benefits all group members to ameliorate the disruptive effects of aggressive disputes by restoring combatants' equilibrium and tolerance. Direct reconciliation among opponents serves one such function, and so may kin-mediated reconciliation. In baboons, victims of aggression appear to accept a friendly signal from their aggressor's close relative as a proxy for reconciliation with the aggressor herself. In order to do so, victims must be able to recognize other females' kinship (or close) relations and make inferences about the intended target of a vocalization. It does not, however, require that they be able to attribute mental states different from their own to others.

Although it seems unlikely that baboons empathize in the sense of projecting their own mental states onto others, the reasons that motivate a female baboon to reconcile with her relative's victim are not immediately obvious. It is easy to postulate why it might be functionally beneficial for females to minimize the disruptive effects of aggression through reconciliation, but the proximate mechanisms for doing so remain elusive. It is unlikely that baboons reconcile with other females' opponents because friendly contact alleviates the anxiety that arises as a consequence of witnessing aggression, because not all bystanders are equally likely to be friendly. Instead, it seems more likely that females reconcile with their relatives' opponents because they identify strongly with their relatives and with their interactions. We take up this idea again in the next chapter.

Despite experiencing grief and anxiety, therefore, baboons appear to have only limited sensitivity to grief and anxiety in others. They maintain strong social bonds and feel bereavement and loss when a close companion disappears. But although they may feel a sort of compassion for others, they do not empathize with them. Just as they do not respond to the contact barks of separated group members unless they themselves are separated from the group, so do they fail to recognize grief or despondency in others unless they too are experiencing loss.

Baboons' theory of mind

Baboons' theory of mind might best be described as a vague intuition about other animals' intentions. Although they do not attribute mental states like ignorance and knowledge to others, baboons and other monkeys do seem to have a rudimentary sensitivity to others' motives and intentions. They seem to understand, for example, that another monkey is likely to be motivated to defend the food that he is looking at, and they appear to use vocalizations to signal and assess intent. We do not yet know whether baboons, like very young children, have an implicit understanding of intentions as mental states or whether they simply recognize other individuals' intentions to behave in a certain way. The distinction between these two interpretations of intention is, in any case, too subtle and indeterminate to distinguish easily through behavioral measures alone. It is difficult to imagine how a baboon could assess whether a signal was being directed toward herself without taking into account a variety of learned behavioral contingencies, including the correlations between call type and behavior and between past and future interactions. There are hints that learned contingencies alone cannot explain all aspects of baboons' behavior, but we cannot yet conclude that baboons regard other baboons—even tacitly—as intentional beings with goals, motives, likes, and dislikes.

Although some experiments suggest that chimpanzees may differ from baboons and other monkeys in their ability to understand other individuals' visual perspectives and the link between seeing and knowing, the evidence is by no means unambiguous or uncontroversial. Chimpanzees' capacity for empathy is similarly disputed. Even if future research does demonstrate unequivocally that chimpanzees recognize others' intentions and even knowledge, what will remain striking is how rarely they appear motivated to share their emotions and goals with others. In the next chapter, we examine similarly unsatisfying and inconclusive evidence for monkeys' and apes' knowledge of self.

NINE

Self-Awareness and Consciousness

It may be freely admitted that no animal is self-conscious, if by this term it is implied, that he reflects on such points, as whence he comes or whither he will go, or what is life and death, and so forth. But how can we feel sure that an old dog with an excellent memory and some power of imagination, as shewn by his dreams, never reflects on his past pleasures or pains in the chase? And this would be a form of self-consciousness.

CHARLES DARWIN, 1871: *THE DESCENT OF MAN, AND SELECTION IN RELATION TO SEX*

What is self-awareness?

Perhaps the only question that has vexed philosophers more than the question of consciousness in humans is the question of consciousness in animals. Darwin's allowance for a "form of self-consciousness" in a dog's reflections on his "past pleasures" was rather charitable, because it did not require the dog to ponder the meaning of life or to be aware of his thoughts. More High Church definitions of consciousness might demand that Darwin's dog also be able to travel mentally back in time and place himself at the scene of the chase. According to this criterion, if we were to ask our dog Eliot to find his ball, it would not be enough for Eliot to remember where the ball is; he should also be able to remember that he, Eliot, placed the ball in that particular bit of shrubbery yesterday.

The ability to attribute mental states like knowledge and ignorance to others would seem to require some level of self-awareness. It is difficult to imagine how a baboon or any other animal could compare her own knowledge with someone else's without some explicit access to her own mind. Given the ambiguous and inconclusive evidence for animals' understanding of others' mental states, however, it is not surprising that evidence for consciousness in animals is also equivocal and patchy.

Operational definitions of consciousness are slippery at best, primarily because—like most mental states—consciousness is not a uniform attribute that one either has or does not have. Philosophers have debated the role and even existence of consciousness in human thinking for millennia and have yet to come up with a satisfying definition. As Daniel Dennett, a philosopher who has wrestled heroically with the problem, puts it, "Consciousness is both the most obvious and most mysterious feature of our minds" (1987a:160; see also Dennett 1991). On the one hand, it seems irrefutable that we are aware of many of our thoughts and emotions and that we are able to introspect about our feelings and desires and plan our future behavior. On the other hand, it is also clear that many of our mental processes remain entirely inaccessible to conscious thought and that many of our actions and perceptions are "unthinking." We know many facts without knowing how we know them, and once we have learned how to perform a particular behavior—even a complex behavior like driving a car—we can repeat the performance without being explicitly aware of every movement.

In struggling with the question of consciousness, William James (1892) suggested that self-awareness was made up of several different components. At the most basic level, there is what he called the "material" or "phenomenal" self, the awareness of one's physical experiences. At a second level, James identified the "social" self, which concerns our awareness of ourselves as distinct individuals, embedded in a group or society that includes many other distinct individuals. At the highest, most complex, level James placed the "spiritual" self, defined as one's "psychic faculties and dispositions" (1892:163). At this level, our thoughts and experiences become available to us for introspection: we can think about what we think, and know what we know. This sort of introspection allows us explicit access to our thoughts, feelings and memories. It has alternately been referred to as "consciousness," "self-awareness," and "metacognition" (Metcalfe and Kober 2005; Nelson 2005).

An organism that is capable of metacognition need not always be aware of his thoughts and knowledge. Indeed, much of the knowledge that we have about ourselves and the world requires no active reflection about how and why we know it. We have no memory, for example, of learning our first language, colors, or numbers. We also often have no memory of how we first acquired even highly specific knowledge, like the fact that there is a giant CITGO sign above the green monster at Fenway Park. We "just know it." In other cases, however, we can subjectively visualize and recall a past event that led to a specific memory: the chair we were sitting in and who was with us when the Red Sox finally won the World Series. Because not all knowledge is associated with a remembered experience, many have found it useful to distinguish between "semantic" and "episodic" memory. Semantic memory concerns memory of facts and events without the added requirement of having to reexperience the place, time, and context in which the memory was obtained. Episodic memory, by contrast, is often referred to as "mental time traveling," because (at least according to most of its current definitions) it requires explicit, subjective awareness of our experiences and of how we acquired particular knowledge and memories (for discussions, see Clayton et al. 2003; Suddendorf and Busby 2003; Tulving 2005; Zentall 2006).

Slippery concepts demand definitions that are ostensibly less so. In this chapter, we will use "consciousness," "self-awareness," and "metacognition" interchangeably, as the ability to introspect explicitly about at least some of our knowledge and beliefs and "to know what we know." We regard episodic memory as a more specialized form of metacognition, because it additionally demands the subjective reexperiencing or anticipation of past or future events.

We do not yet completely understand why some of our thoughts become conscious to us, and the degree to which metacognition helps to coordinate other mental processes also remains unclear. Nevertheless, there does seem to be agreement that most of our mental processes remain unconscious and largely inaccessible, and that there is no reason why theories about our own minds should be any less fallible than theories about the minds of others.

The function of self-awareness remains elusive. Humphrey (1986) speculated that consciousness has evolved to allow us to predict the behavior of others on the basis of introspection about our own motives, thoughts, and beliefs. So, for example, a baboon might predict that she can retaliate effectively against an opponent by threatening her opponent's kin because she knows that similar retaliation against herself would make *her* angry. Self-awareness, in this view, is a necessary precursor to speculating about the minds of others. Interestingly, the medial prefrontal cortex, which appears to play such an important role in the assessment of others' mental states, shows similar activation when people consider their *own* thoughts and beliefs (Frith and Frith 2003; Saxe et al. 2004; Mitchell et al. 2005). This similarity is just what we would expect if introspection served as a template for inferences about others' thoughts.

Explicit access to our experiences and memories may also assist us in planning for the future. Adults with retrograde amnesia often know and remember many things—their semantic memory is intact—but they have no memory of anything that happened specifically to them. They cannot bring into conscious awareness any event that they have witnessed or experienced and they cannot plan their future actions (Tulving 2005). In contrast, the mental time traveling associated with episodic memory allows us not only to reexperience past events but also to anticipate ones that have yet to arise (Suddendorf and Busby 2003; Zentall 2006). By drawing on past experiences, we can simulate hypothetical future social interactions and imagine how we might improve upon a difficult social negotiation, tool, or hunting technique. Explicit introspection about our mental states, in other words, may be essential for planning about how to manipulate both other individuals and objects.

Children's self-awareness

Like children's theory of mind, children's awareness of self develops gradually, and their manifestations of self-identity vary substantially with age. Most children recognize their mirror images by around 18 months of age (reviewed by Courage et al. 2004), at about the same age that they begin to experience emotions like embarrassment and shame (Barrett 2005). Very young children can also reliably identify their own place in a family or social group (Damon and Hart 1982). Not until several years later, however, will they begin to recognize that their own knowledge and beliefs can be different from someone else's and understand how they acquired a particular memory or knowledge.

Episodic memory emerges in children at roughly four years of age—around the same age as complex mental state attribution—suggesting that the ability to separate one's own experiences, beliefs, and thoughts from others' is crucially linked to an explicit sense of self (Atance and O'Neill 2005). Although younger children can learn and remember many things without much difficulty, they are often unable to recount how they came to know them (Nelson 2005). For example, when three-, four-, and five-year-old children were shown the contents of a drawer, all of the children could later recall the objects they had seen. Only the older children, however, could explain how they knew what was in the drawer ("You showed us!"). The younger children stated that they had always known what was in it, or that they "just knew" (Gopnik and Graf 1988; see also Taylor et al. 1994; Drummey and Newcombe 2002). In this sense, young children manifest what Nelson (2005) has called "childhood amnesia": they know many things, but they cannot remember how or why they know it. The lack of such explicit, experiential memory limits children's ability to imagine themselves in hypothetical scenarios and to reexperience events or emotions that are associated with a particular memory.

Investigating self-awareness in animals

The question we pose in this chapter, then, is not whether animals remember events, social companions, or the location of food—clearly they do. It is, instead, whether they know what they know, and how and why they know it. Because such explicit access to memory demands introspection and the ability to separate one's own beliefs and knowledge from others', it may be beyond the capacity of any animal. Perhaps, as John Donne wrote in a sermon in 1628, "The beast does but know, but the man knows that he knows" (quoted by Kinsbourne 2005:144).

In our daily interactions with others, we take the question of human consciousness for granted. We are aware of at least some of our own states of mind, and we can use this awareness to predict both our own and other individuals' behavior. Through introspection into our own thoughts, we assume that others are also aware of their own thoughts and aware of their identities as individuals distinct from all others. These intuitive theories about our own and other people's mental states are much more difficult to apply to animals. In the absence of language, it is difficult if not impossible to ask someone how he knows what he knows. Although we can ask a human whether he is reexperiencing the particular event when he recalls a specific memory, such interviews are clearly impossible in the case of animals. Simply asking Eliot to find his ball will not allow us to determine whether he "just knows" that the ball is in the bush to the right of the door or whether he knows this because he can recall putting the ball there.

As mentioned earlier, an explicit sense of self emerges in children at roughly the same age as the ability to attribute knowledge and beliefs to others. Indeed, it is hard to imagine how it would be possible for a person to recognize that another individual has a belief or knowledge different from his own without having at least some conscious access to his own beliefs and knowledge. And because there is little evidence that any animal is capable of this sort of complex mental state attribution, it may also be the case that animals are not capable of metacognition and the mental time traveling involved in episodic memory. Nonetheless, as we have seen, baboons and other animals are clearly able to distinguish their own social relationships from those of others, and there is some suggestion that they may recognize other individuals' intentions and motives. It therefore seems possible that they might also have some limited access to their own knowledge and some ability to plan their future behavior.

Baboons often behave as if they were planning and rehearsing a social interaction. We once watched Margaret, Sylvia's juvenile granddaughter, sit for over half an hour, chewing slowly on a palm nut and staring at Selo, the group's alpha female and doyenne (Fig. 39). Selo had a young infant, and Margaret seemed to be debating whether and how she might approach to inspect the baby. But was Margaret imagining dramas starring herself, perhaps as the heroine who saves the baby from the jaws of a leopard to the eternal gratitude of a tearful Selo? Or was she considering possible strategies: "If I grunt, will she let me approach?"

Figure 39. The juvenile female Margaret considers whether to approach Selo. Photograph by Anne Engh.

Even this thought process would seem to demand some form of conscious reflection. Or maybe Margaret was hesitating to approach because she was experiencing feelings of uncertainty, based on memories of previous interactions with Selo. Would such feelings of uncertainty also demand some degree of introspection? Or was Margaret just trying to crack the nut in her mouth?

Below, we review some attempts to examine the question of self-awareness in animals. Because we were unable to address this question directly with baboons, much of the evidence we describe comes from work on captive animals. We first describe this work and then return to a discussion of James' social self as it applies to baboons.

The phenomenal self

Most of our body's basic needs are regulated by physiological mechanisms that are not consciously accessible. Other sensory experiences, like a pinprick or a visual image, can be experienced consciously. It seems probable that almost all animals have an elementary recognition of their phenomenal, or material, self, in the sense that they react to painful stimuli and distinguish between sensory inputs that come from their own bodies and sensory inputs that come from elsewhere. With some exceptions, however, most animals' sense of their material self seems to remain tacit, without the individual being actively aware that he can see or alter his unique self.

Tests of mirror self-recognition, originally devised by Gallup (1970), offer strong evidence that apes, but not monkeys, recognize that the face they see in the mirror is their own (see also Heyes 1994; Tomasello and Call 1997). Although it is not altogether clear what aspects of consciousness are reflected by tests with mirrors, they do reveal a consistent and qualitative difference between apes and monkeys, and they suggest that apes have some capacity to make material self-recognition at least partially accessible to thought. Nevertheless, although monkeys may not recognize their mirror images as themselves, they do appear to be at least partially aware of their perceptions. The phenomenon of "blindsight" illustrates this point.

We sometimes lose conscious access to phenomena of which we are normally aware. For example, humans with damage to the visual cortex experience blindness, in the sense that they are unaware of seeing. Often, however, they can still unconsciously detect, locate, and discriminate visual events in their blind field even though they report that they are unable to do so (Weiskrantz 1998). This occurs because the eyes send pathways not just to the striate cortex, where images become consciously accessible, but also to other areas in different subcortical parts of the brain. As a result, people with blindsight experience a dissociation between their awareness of seeing and their perceptual capacity to do so. Conscious introspection of visual information is lost. When asked to state whether a pattern of light is vertical or horizontal, people with blindsight are highly accurate in their answers, but they say that they see nothing and are only guessing.

Similar results have been obtained from monkeys with blindsight. Stoerig and Cowey (1995) removed part of the striate cortex in the left hemisphere of four macaques, so that the monkeys were "blind" in their right eye. The monkeys were then trained in a simple discrimination task in which they had to touch one of five buttons that was briefly lit. The monkeys also received blank trials, when no button was lit. After these, the monkeys had to press another, "no light," button to signal that they had seen nothing. All of the monkeys accurately touched the lit button, even when it was presented in their "blind" visual field. However, after these trials, the monkeys also pressed the "no light" button, effectively reporting that they were not aware of seeing an event that they had clearly processed subcortically.

Metacognition

Monitoring knowledge

Monkeys with blindsight seem to have some awareness of what they think they see. Other recent experiments suggest that, Donne's remarks notwithstanding, monkeys may have a rudimentary ability to evaluate what they do and do not know.

Consider the state of uncertainty. When asked to identify a person whom we have met only once, we often experience a sense of some indecision. In so doing, we are making a tacit assessment of our knowledge. Similarly, when a dog is asked to jump into the back of a car, he will often whine, hesitate, and run around in circles before attempting (or refusing) the jump. For the psychologist Edward Tolman (1932:206) such "running, or looking, back and forth" hinted at mental turmoil and "constitutes a conscious awareness." However, it is entirely possible to feel uncertain and hesitant without being actively aware of these feelings. True metacognition would seem to require more explicit introspection and deliberate evaluation of one's knowledge. As Tolman put it, introspection "requires that [the individual] can report that he is thus adjustmentally running-back-and-forth" (1932:241).

Hampton (2001, 2005) investigated two rhesus macaques' ability to make prospective judgments about their knowledge. Each monkey first saw a picture displayed on a computer screen for several seconds. After a variable delay, the monkey was given a forced recognition test. The original picture was displayed on the screen along with three novel ones, and the monkey had to pick the picture he had seen earlier. On other trials, the monkeys were not forced to take the test, but given the choice of escaping it. If they chose to take the test, they received a large reward if they were correct, but none if they were wrong. If they chose to escape, they always received a small reward. The monkeys performed better on tests that they were free to take than on ones that they were forced to take, suggesting that the monkeys were, at some level, evaluating their knowledge prior to deciding whether or not to take the test.

Similarly, Son and Kornell (2005) devised a series of experiments examining rhesus macaques' confidence judgments about their *prior* decisions. In one experiment, the monkeys sat in front of a computer screen that displayed eight squares containing a variable number of items. Their job was to choose the square with the fewest number of items. In some trials, this choice was easy, because the correct square contained noticeably fewer items than the others (say, 1 vs. 8). In more difficult trials, the correct square contained only one or two fewer items (say, 6 vs. 7). Both monkeys became skilled at the task.

Each monkey was then taught to wager bets about the accuracy of his choice (Fig. 40). After the monkey had made his choice, the squares disappeared from the computer screen and were replaced by two symbols, one representing a high-risk bet, the other representing a low-risk bet. If the monkey placed a high-risk bet and his choice had been correct, three tokens dropped into a tube displayed on the screen. But if he wagered a high-risk bet after an incorrect choice, three tokens flew out of the tube—to the monkey's blatant displeasure and distress. In contrast, when the monkey made a low-risk bet, one token was added to the tube regardless of whether his previous choice had been correct or incorrect. After the accumulation of six tokens, the monkey received a food reward. Although it took more than a year and a half for the monkeys to become proficient with the betting paradigm, both eventually learned to place bets. Furthermore, both monkeys were more likely to choose the high-risk bet after they had chosen correctly than after they

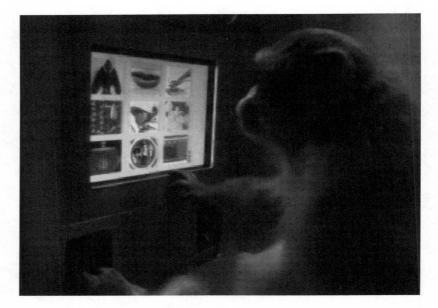

Figure 40. The rhesus macaque Ebbinghaus attempts to decide which picture he has seen previously. The tube holding the betting tokens is at the bottom right of the screen. Photograph by Lisa Son.

had chosen incorrectly, suggesting that they not only remembered their previous decision but also were able to assess their confidence in it. Son and Kornell were able to rule out the possibility that the monkeys were simply relying on cues such as response time when deciding whether to make a high- or low-risk bet, and the monkeys generalized the betting paradigm easily to new tests, including ones based on their memory of previously observed pictures (see also Shields et al. 1997, 2005; Smith et al. 1995, 2003; Smith 2005; Washburn et al. 2006 for similar experiments with monkeys and dolphins).

A number of experiments, then, have suggested that monkeys are able to access and monitor their knowledge. Whether or not they are explicitly aware of doing so, however, remains unclear. Although the monkeys in these experiments might have been able to introspect about why they felt uncertain, it seems more likely that they just felt hesitant without being aware of the causes of their uncertainty. In the latter case, the monkeys might simply have learned to avoid high-risk options (and the possibility of losing a reward) whenever they felt uneasy. Even this lower, implicit metacognition, though, would provide monkeys with an accurate means to assess their knowledge. Indeed, as Son and Kornell point out, many of our own certainty judgments are also implicit and "unthinking." In game shows, for example, contestants are often required to press a buzzer to signal whether or not they know the answer to a question. These metajudgments are often made very quickly, before the contestants have retrieved the answer. Nevertheless, contestants' judgments about the accuracy of their yet-to-be-accessed knowledge is usually correct. Despite humans' ability to introspect about what we know, therefore, we are not always aware of why we feel certain or uncertain. Whether or not monkeys are aware of their uncertainty is, as Son and Kornell state, "endlessly arguable."

It is difficult to imagine how the elegant experiments just described could be transferred to free-ranging animals. Baboons often act as if they are reassessing their knowledge, but whether or not they are explicitly aware of doing so is impossible to determine. The anecdotes are intriguing, if inconclusive. To wit: it was an impossibly hot day and the baboon group was spread out over a large distance as it traveled to a distant woodland. As a result, few baboons had witnessed Nicky, the third-ranking male, topple Morgan from his second-ranking position. There had been a brief but decisive scuffle over a clutch of eggs, and Morgan had been forced to cede both the eggs and his rank to Nicky. Several hours later, one of us found ourselves sitting under an acacia bush with Third Man, the fourth-ranking male in the group. The group was slowly heading out across an open plain, but we were both reluctant to move into the sun. About 50 meters away, Nicky approached Morgan and Morgan gave way. Third Man, who had appeared to be dozing, started, sat upright, and grunted. Apparently, he had not known about the switch in ranks and was surprised by what he saw. It is certainly tempting to conclude that the incident provoked a degree of uncertainty in Third Man, forcing him to admit that his knowledge of the male dominance hierarchy needed to be updated.

Similarly, before initiating a water crossing (or leaving the sleeping site) baboons spend a long time dithering on the shore. Certain adults seem to take one of two roles. The "initiators" make the initial forays into the water. Usually, these are older males or females of varying ranks. As other group members watch and grunt, the initiator wades across the first body of water to a nearby small island or termite mound. He or she then sits and waits to be joined by others, gazing back at the onlookers and exchanging grunts with them (Fig. 41). Although many other group members may eventually follow the initiator's lead, it is not until one of the "deciders" enters the water that the full crossing is made. Deciders are most often the adult female members of the highest-ranking matri-

Figure 41. How do baboons decide when to make a water crossing? Photograph by Keena Seyfarth.

line, and Selo is the most persuasive decider of them all. Unless she joins the crossing, the initiators almost always return. When the initiator is a subordinate female, Selo will sometimes approach that female after the first crossing has been made and exchange grunts, lipsmacks, and hugs with her.

These observations beg more questions than they answer. Indeed, the puzzle about who initiates and decides the direction of group movements has vexed primatologists for decades (e.g., Kummer 1968; Boinski and Garber 2000). It often seems that other group members are following the lead of specific individuals, but these individuals are not necessarily the ones that initiate the movement. Why do some baboons act as initiators? Do they know that they are more experienced at making water crossings, or do they simply feel less afraid (and more impatient) than others? Does Selo know that she is the ultimate arbitrator in the decision? Does she assess her level of certainty before deciding to join the crossing? To the human observer, it seems very much as if the grunts, lipsmacks, and looks exchanged between initiators and onlookers function as mutual acknowledgments of risk and the need to coordinate.

The baboons appear to be aware of each other's intentions and levels of uncertainty.

At the same time, however, it may be equally (if not more) probable that other group members follow Selo's lead simply because they do not want to become separated from the rest of the group, and Selo, being the highest-ranking and a member of the group's largest family, is more likely to be followed by more animals than other females. As a result, Selo may have acquired the role of "decider" without ever explicitly knowing how or why she has done so. As yet, we have been unable to devise a way to test between these two explanations.

Episodic memory

Many food-caching birds and rodents remember not just where they have stored food but also which sites they have already depleted. If a nutcracker's retrieval session is interrupted he will avoid previously depleted sites in his next retrieval session, even if many days elapse before he can resume his search (Balda and Kamil 1992). By following this "win-shift" strategy, the nutcracker demonstrates that that he remembers not only which sites still contain seeds but also which sites do not. But when he remembers which sites are now depleted, does the bird also remember that he was the one that depleted them?

Experiments conducted with scrub jays suggest that food-caching birds remember not just where they stored food but also what kind of food they stored and when they stored it (Clayton and Dickinson 1998). Given the opportunity to retrieve previously stored peanuts or mealworms, scrub jays preferentially retrieve the more desirable mealworms, but only if they stored the worms recently, before they have had time to decay. After 124 hours' delay, the jays choose to retrieve peanuts. This behavior has led Clayton and colleagues to argue that scrub jays have an "episodic-like" memory because they appear to remember what they did, when they did it, and where they did it (see also Griffiths et al. 1999; Clayton et al. 2003; de Kort et al. 2005).

Scrub jays also seem to use memories of their own behavior and experiences as a template for predicting the behavior of others. Recall that scrub jays and ravens are more likely to recover and rehide previously stored food when being observed by a potential pilferer than when alone. In scrub jays, birds that have learned to pilfer from others are more likely to rehide their caches than are naïve birds that have not yet learned to steal (Emery and Clayton 2001, 2004). The birds' hiding strategies seem to be influenced by their expectations of being robbed, which in turn depends on their experiences as thieves.

Memory of recent actions is not restricted to food-storing birds. Rats accurately remember what they did and when they did it (Ergorul and Eichenbaum 2004), and trained dolphins can report not only what they have just done, but also what they have *not* just done (Mercado et al. 1998). In both birds and mammals, the hippocampus plays a crucial role in these episodic-like memories (Eichenbaum et al. 2005), enabling individuals to review their recent actions (Foster and Wilson 2006).

There is little doubt that scrub jays, rats, and other animals are capable of remembering what, where, and when something occurred. They use their memories flexibly to plan future actions—for example, where next to retrieve food. What remains unclear is whether they are capable of the sort of metacognition that would allow them to recognize how they obtained their information. If we could interview them, they might simply report—like three year-old children—that they "just knew" where and when particular food items had been stored. Similarly, although scrub jays' memories of their own behavior may guide their responses to other birds, there is no evidence that they are engaged in mental time-traveling when they do so—that they can explicitly recall the meal worm that they pilfered from Freddy last Tuesday. In the absence of evidence that scrub jays can actively reflect about the source of their knowledge, their memory cannot be described as truly episodic.

But this may be unfair. In the absence of an interview, how can we ascertain with any certainty whether or not *any* animal has episodic memory? Apes, too, often act as if they have episodic memory. In one experiment, Mulcahy and Call (2006) trained a bonobo (or pygmy chimpanzee, *Pan paniscus*) and an orangutan to use a particular tool to open a box containing food. In test trials, each individual was presented with a choice of eight different tools, only two of which were suitable for opening the box. The ape could choose to select a tool or not. After five minutes, she was ushered from the testing room and sent to her sleeping room. The remaining tools were then removed from the room. Fourteen hours later, she was allowed back into the testing room. If she had chosen the correct tool the day before, she could then open the box and retrieve the food. Neither of the subjects chose a tool in the first trial. In most of the next 11 trials, though, they not only chose the correct tool before leaving the testing room but also remembered to bring it back with them when they left their sleeping room the next morning. The apes did not simply learn to carry the tool in and out of the testing room because they associated tool transport with a reward, because when the experimenters removed the box and rewarded the subjects just for returning the tool to the testing room, the apes typically neglected to bring it back. Instead, by choosing the correct tool many hours before they would receive a reward, both subjects appeared to be planning for the future.

But even this impressive behavior is subject to some caveats. When the apes chose a specific tool did they imagine themselves using the tool tomorrow? Or had they just learned that they had to take the tool with them when they left the room? In the wild, chimpanzees show evidence of planning when they select a prodding stick before going to a termite mound or collect nuts and stones before heading to an anvil. When selecting a stick, do they imagine themselves prodding the stick into the mound and licking off delicious termites, or do they "just know" that they need to find a good stick before they begin to fish? Just as monkeys might experience feelings of uncertainty without reflecting explicitly about why they feel that way, so might chimpanzees show considerable foresight without explicitly projecting themselves into the future.

The social self

Children's awareness of their own identities as unique individuals develops at roughly the same age as their awareness of the unique identities of others (e.g., Damon and Hart 1982; Rotenberg 1982; Gopnik and Meltzoff 1994). For James (1892), a crucial component of consciousness was the awareness of oneself as a distinct individual, a member of a social network comprised of other distinct individuals. We define our social selves in part by reference to others; there cannot be an "I" without a "you" or a "they" for comparison.

We have argued that a baboon would be in a state of complete nervous exhaustion if she responded to every vocalization or behavior she heard or saw. The crucial component of the social self, therefore, is the ability to recognize when a vocalization, gaze, or other communicative behavior is directed *at you* and not at someone else. As we described in Chapter 8, baboons seem very sensitive to the contingencies surrounding social interactions and are masters at determining whether they or someone else is the target of a call. A baboon would not be able to decide "Who, me?" without also being able to determine "Not me."

When a vervet monkey observes a fight between one of her own close relatives and another individual, she often retaliates by threatening a member of her relative's opponent's family (Cheney and Seyfarth

1990). In so doing, she behaves as if she recognizes not only the relationships that exist within her opponent's family, but also the relationships that exist within her *own* family. Monkeys also seem to recognize their own dominance ranks. When a dominant female vervet or baboon approaches two lower-ranking females who are grooming, it is almost invariably the lower-ranking of the two groomers who moves away (Chapter 6). Both the female who leaves and the one who stays seem to recognize their own status relative to each other.

Baboons' stress responses are exquisitely sensitive to the distinction between "I" and "you." When a male baboon begins to rise in rank, other high-ranking males experience an elevation in glucocorticoid levels. Low-ranking males, in contrast, show no such response. Instead, their glucocorticoid levels are higher during periods when the male dominance hierarchy is stable (Chapter 4). These differences suggest that males' stress responses are influenced by the loss of social control. During periods of rank stability, high-ranking males enjoy predictable and privileged access to food and estrous females, while low-ranking males do not. In contrast, when the male dominance hierarchy is unstable and the dominance positions of high-ranking males are in jeopardy, dominant males experience a loss of social control, and a concomitant increase in glucocorticoid levels (Sapolsky 1992, 1993; Bergman et al. 2005). The males' stress responses occur as a result of events that are happening to *them*, and not to others (Fig. 42). Similarly, female baboons show stress responses primarily to events that affect *themselves*: the loss of a close relative, an infanticidal threat to their own infant, and rank instability that affects their own dominance position (Engh et al. 2006a,b).

Although baboons and other monkeys may behave as if they recognize their unique place in the social network and distinguish between events that affect them and those that do not, they may not be explicitly aware of doing so. Just as the stress response does not depend on introspection, so might baboons not reflect actively on their rank positions and familial relationships.

As we have described, female baboons often show friendly behavior toward individuals whom their close relatives have recently threatened. Why should a baboon feel motivated to reconcile on behalf of her relatives? Empathy seems unlikely, because baboons appear unable to attribute mental states different from their own to others.

Instead, kin-mediated reconciliation may occur because baboons identify so strongly with their close relatives that interactions involving family members are viewed as surrogates for interactions involving

Figure 42. Spock searches for his consort female, Selo. Males' glucocorticoid levels show a temporary rise when they are engaged in sexual consortships. Photograph by Roman Wittig.

themselves. According to this explanation, Selo's bond with her daughter Palm is so strong that she regards a fight involving Palm and Comet as tantamount to one involving herself and Comet. She is therefore motivated not only to form alliances with Palm but also to reconcile with Palm's victims. For Selo, Palm's interactions become *her* interactions. And if reconciliation functions to restore equilibrium between opponents, Selo should be as likely to reconcile with Palm's opponents as with her own. In contrast, fights involving nonrelatives do not evoke the same sense of self-involvement. This sense of self-involvement is probably not as visceral and instinctive as the one that motivates a mother animal to defend her offspring, because the behavior is not immediately linked to reproductive success. Baboons choose whether or not to form an alliance with a relative who is already winning a dispute, and they choose whether or not to reconcile with an individual who has already lost a fight to their relative. They do not always do so; the response is not reflexive.

Baboons' nepotistic system of rank acquisition and maintenance ensures that a rank reversal involving one family member is likely to expand to engulf all other family members. It therefore behooves a female to identify strongly with her relatives' altercations and disputes. In between-family fights, the baboon's "I" expands to include all of her close kin; in within-family fights, it contracts to include only herself. This explanation serves for baboons as much as for the Montagues and Capulets. In both cases, moreover, the urge to equate one's self with one's family is probably largely unconscious.

The suggestion, then, is that a baboon can locate not only others but also herself within the group's social network. To do this, she would seem to need some image of herself as a unique social being, distinct from all others and characterized by a unique set of social relationships with particular others. This is not the kind of self-consciousness that will ever give rise to a full-blown identity crisis, but it is a necessary step along the way to a truly human consciousness, of not only one's social position but also one's thoughts and beliefs.

Thus, events that affect us through our families are seldom regarded with as much equanimity as those that affect others. Jane Austen noted this unsurprising phenomenon long ago in *Pride and Prejudice,* when she had Mr. Bennet remark, "For what do we live for, but to make sport for our neighbours, and laugh at them in our turn?" But when Lydia runs off with the nefarious Wyckham, all bemused aplomb is abandoned. Elizabeth Bennet recognizes immediately that Lydia's behavior has destroyed the reputation of the entire family. As Mr. Collins sanctimoniously writes, "This false step in one daughter will be injurious to the fortunes of all the others, for who ... will connect themselves to such a family?" Elizabeth observes her neighbors' feigned concern and laments "Assistance is impossible; condolence, insufferable. Let them triumph over us at a distance, and be satisfied." Events that happen to *you* can be regarded with curiosity, amusement, and even gloating; those that happen to *me* are far more sobering.

TEN

Communication

Their speech is the only gateway to their minds, and through it we must pass if we would learn their secret thoughts and measure the distance from mind to mind. **RICHARD GARNER, 1892:** *THE SPEECH OF MONKEYS*

When you barke, doe it with judgement.
SCHOOLMASTER TO THE BABOON, BAVIAN, IN WILLIAM SHAKESPEARE AND/OR JOHN FLETCHER, 1614: *THE TWO NOBLE KINSMEN*

In their responses to our playback experiments, baboons reveal their thoughts. A baboon who ignores the sequence "Sierra threat-grunts and Luxe screams" but responds strongly when she hears "Luxe threat-grunts and Sierra screams" tells us that she knows who is calling, what is occurring, and that Sierra should outrank Luxe. Her aptitude for deducing a rich narrative from a stream of sounds is impressive. It also makes her small repertoire of relatively stereotypic calls seem all the more paradoxical. Why should a monkey who can extract nuanced information from others' vocalizations be unable to convey equally nuanced information in her own?

Since Aristotle, philosophers and scientists have accepted the proposition that human language and thought are intimately related and that language, being public, can help us understand thought, which would otherwise be private. But the same argument has rarely been applied to animals, probably because human language and animal communication have always seemed so different. Darwin's views were typical of his time. He believed that the production of sounds by animals originally appeared as the involuntary consequence of other bodily movements:

When the sensorium is strongly excited, the muscles of the body are generally thrown into violent action; and as a consequence, loud sounds are uttered, however silent the animal may generally be, and although the sounds may be of no use. (Darwin 1871/1981:83).

Darwin also noted that over evolutionary time the production of some sounds had come to be associated with specific emotions, such as pain, pleasure, or rage, and, as a result, these vocalizations had come to serve a communicative function. The roaring of lions and the growling of dogs signal these animals' rage and "thereby endeavour to strike terror into their enemies." The incessant calling of males in the breeding season signals their "anticipation of the strongest pleasure which animals are capable of feeling" and thereby "endeavours to charm or excite the female" (Darwin 1871/1981:84–85).

More than a century later, little had changed. In 1986, after years of field research, Jane Goodall concluded that "chimpanzee calls are, for the most part, dictated by emotions" (Goodall 1986:125), while in 1990 the linguist Derek Bickerton stated that primate vocalizations are "quite automatic and impossible to suppress" (Bickerton 1990:142). Like Darwin before them, Goodall and Bickerton drew a sharp distinction between the learned, voluntary sounds that are used in human language and the innate, reflexive sounds that are used in animal communication. Whereas language is a semantic system in which words can represent thoughts, actions, and events in the world, animal communication is nothing more than the expression of emotions. And because animal calls are reflexively linked to emotions, the close relation between a speaker's vocalizations and his thoughts—so obvious in human language—simply does not exist in animals.

At the same time, many scientists and philosophers have also conceded that, in their comprehension of human signs and speech, animals often seem remarkably similar to humans. Indeed, two pages after arguing that animal signals are nothing more than reflexive reactions, Darwin wrote:

That which distinguishes man from the lower animals is not the understanding of articulate sounds, for, as every one knows, dogs understand many words and sentences. In this respect they are at the same stage of development as infants, between the ages of ten and twelve months, who understand many words and short sentences, but cannot yet utter a single word ... Nor is it the mere capacity of connecting definite sounds with definite ideas; for it is certain that some parrots, which have been taught to speak, connect unerringly words with things, and persons with events. (Darwin 1871/1981:85)

For Darwin and many others, then, there was a huge disconnect in animals between the mental processes underlying call production and those underlying call comprehension. This view has much validity. Dogs, for example, can learn to associate different words with specific toys (ball) and events (car ride), even though their barks are relatively invariant and stereotypic. The apparent discontinuity between production and comprehension results in an odd imbalance: dogs, monkeys, and other animals can learn many sound-meaning pairs but cannot produce new words. They understand conceptual relations but cannot attach labels to them (see also Chapter 11; Cheney and Seyfarth 1998).

But there are at least three flaws to the proposition that there is a vast, yawning gap between the mental mechanisms that underlie call production and those that underlie comprehension. First, listeners are also signalers. The vervet monkey who responds to a companion's eagle alarm call by looking up into the sky will on some other occasion be the individual who is giving the eagle alarm call. It therefore seems unlikely that the mental representations that accompany the interpretation of an eagle alarm call are always qualitatively different from those that accompany its production. We discuss this point further in Chapter 11.

Second, even if animal vocalizations were simply innate, unthinking reflexes (and we believe that they are not), they still have the potential to carry rich semantic meaning. To understand why, consider an allegory borrowed from David Premack (1975), who was perhaps the first to point out that a nonlinguistic system of communication based entirely on emotion can effectively become semantic. Suppose, Premack argued, you know that I love strawberries more than anything else. And you also know that more than anything else I hate and fear snakes. One day, when I am out of sight behind a bush, you hear me give a great shout of joy. If you know that I only give this call when I've found strawberries, my cry tells you unambiguously that there are strawberries behind the bush. I might just as well have said the word "strawberries." Similarly, if you hear a scream and can be certain that I only give this call to snakes, my scream tells you unambiguously that I have seen a snake. The moral is: whenever a listener can detect a predictable pattern in another's vocalizations, even a system of communication based entirely on emotions can become one that conveys information about objects and events in the world.

Third, monkeys and other animals do not just respond reflexively to stimuli in their environment. To the contrary, the causal factors underlying call production are myriad and complex, and there is often a close relationship between a particular call type and a specific object or event in the external world. The crucial analytical method is to take each communicative event and deconstruct it by looking separately at the signaler and the listener—at the factors that cause one individual to vocalize, and the information that a listener can acquire when he hears a call.

What causes an animal to vocalize?

Alarm calls

Some animal vocalizations are elicited by a broad array of stimuli, while others are much more specific. Both suricates (a South African mongoose, *Suricata suricatta*) and Diana monkeys (a West African forest monkey) give alert calls in response to many different stimuli, including mammalian and avian predators, large nonpredatory animals, falling trees, and social disturbances within the group (Gautier and Gautier 1977; Zuberbuhler et al. 1997; Manser 2001). When animals hear an alert call they look intently in the direction of the signaler, as if searching for more specific information. These alert calls stand in marked contrast to the alarm calls that suricates and Diana monkeys give in response to specific types of predator. Suricates give one alarm call type to mammalian predators (primarily jackals, *Canis mesomelas*), a second type to hawks and eagles, and a third to snakes (Manser 2001). Listeners respond in qualitatively different ways to the different alarm call types. Upon hearing a mammalian predator alarm, they run to the nearest burrow, upon hearing an eagle alarm they scan they sky, and upon hearing a snake alarm they approach the sound, giving alarm calls themselves. Similarly, Diana monkeys give acoustically distinct alarm calls in response to mammalian predators like leopards and to avian predators like the crowned eagle (*Stephanoaetus coronatus*). When they hear a leopard alarm call they climb higher in the trees; when they hear an eagle alarm call they run down, out of the forest canopy (Zuberbuhler et al. 1997, 1999).

In many species that give different alarm calls to different predators, variation in predator type is the primary stimulus that determines which alarm call is given. Variation in other aspects of the social and ecological context is relatively unimportant. Like Diana monkeys, vervet monkeys have several specialized alarm calls. One is given to mammalian carnivores like leopards, another to large raptors like the martial or crowned eagle, and a third to snakes (Struhsaker 1967). When vervets on the ground hear a leopard alarm call they run into the trees. When they hear an eagle alarm call they look up into the air, and when they hear a snake alarm call they inspect the bushes, trees, and grass around them (Seyfarth et al. 1980). For both Diana and vervet monkeys, alarm calls are truly predator-specific: the sight or sound of a leopard or eagle elicits the appropriate alarm call regardless of predator numbers, distance, or elevation (Seyfarth et al. 1980; Zuberbuhler 2000).

In other cases, the relation between the eliciting stimulus and alarm call type is more complex. In suricates, signalers also vary the acoustic properties of each alarm call type in a manner that is associated with levels of urgency. They give low urgency mammalian predator alarms to distant jackals and high urgency mammalian predator alarms to closer ones. The suricates also produce low and high urgency versions of their eagle and snake alarms (Manser 2001). Within each alarm call category, they respond more strongly to high urgency than to low urgency variants (Manser et al. 2001a). The eliciting stimuli for suricate alarms, therefore, include both predator type and some features of the immediate context that are correlated with the caller's perception of urgency (Manser et al. 2001b).

The alarm calls of vervet monkeys, Diana monkeys, and suricates function, like words, to designate different predator types. In other species, alarm calls communicate information about urgency but not about the nature of the threat. For example, if a predator arrives suddenly and there is little time to escape, California ground squirrels (*Spermophilus beecheyi*) give whistle alarms regardless of whether the predator is a mammal or a raptor. When a predator is spotted at a distance, they give chatter-chat alarms, again regardless of predator type (Owings and Hennessy 1984; see also Blumstein and Armitage 1997).

Martial eagles and crowned eagles—the only African eagles big enough to prey on baboons—are rare or absent in the Okavango. As a result, we do not know whether baboons, like vervets and Diana monkeys, might have an eagle-specific alarm call. Baboons do, however, give alarm calls to mammalian carnivores like lions and leopards, as well as to crocodiles and snakes. Males produce loud alarm wahoos and females and juveniles give a distinctive alarm bark (Fig. 43). Alarm wahoos are acoustically similar to the contest wahoos that males give

Figure 43. Baboons give alarm calls to the lions that have treed them. Photograph by Chris Harvey.

during competitive contests with other males (Chapter 4), but the two types of wahoo differ according to a number of acoustic measures (Fischer et al. 2002). Similarly, the alarm barks given by females and juveniles are acoustically similar to the contact barks that baboons give when they become lost or separated from their companions (Chapter 8). Again, however, there are subtle acoustic differences between the two bark types that allow them to be distinguished by ear (Fischer et al. 2001a).

Male and female baboons give a slightly different alarm bark in response to crocodiles and snakes. Although we have not been able to study this vocalization systematically, it is our strong impression that listeners distinguish between alarm barks given to mammalian carnivores and those given to crocodiles. When the baboons are foraging near water, where attacks by lions and crocodiles are equally likely, they respond to "lion" barks by running toward trees and to "crocodile" barks by running a short distance from the water before stopping to peer into it. In the absence of playback experiments, though, we cannot confirm this impression.

Other vocalizations

Unlike predator alarm calls, which depend in a fairly simple way on the type of predator or the degree of danger, the vocalizations given by animals during social interactions are elicited by a more complex array of factors that may include both the immediate social context and the history of interactions between the particular individuals involved. Baboon grunts offer an excellent example.

As we mentioned in Chapter 5, grunts are the baboons' most common vocalization. They are individually distinctive and given in a variety of nonaggressive circumstances. While all baboon grunts sound similar to us, they are actually of two acoustically graded types (Owren et al. 1997; Rendall 2003). One, the move grunt, is typically given in the context of group movement, either as the animals forage through woods or tall grass or when a group move is just beginning. A female may stand up, look at several other individuals, and give one or two move grunts. This attracts the attention of others, who may answer her with a move grunt of their own. Move grunts occur at particularly high rates when the move is potentially dangerous—for example, during water crossings. Like similar calls given by other primate species, move grunts function to alert other individuals to the signaler's intentions to travel in a particular direction. This is important, because in large social groups there may be conflicts of interests about where or whether to move, but strong incentives for group members to remain together (Silk in press a).

The second, slightly different infant grunt is given by adult females during many sorts of friendly interactions and provides information about the signaler's disposition (see Chapters 5 and 8). Infant grunts are most commonly given in the context of infant handling, but they also occur during grooming, other friendly behavior, and reconciliation (Fig. 44). The two grunt subtypes thus differ in the specificity of the

Figure 44. The juvenile female Domino grunts as she examines the infant of CP, a lower-ranking female. Photograph by Keena Seyfarth.

stimuli that elicit them. Move grunts are linked to a specific context. By comparison, infant grunts are given in a wide variety of friendly situations. Whereas infant grunts function to signal benign intent toward one specific individual, move grunts broadcast the signaler's intentions to many individuals.

Move and infant grunts exemplify the variation found in the baboon vocal repertoire, and indeed in the vocal repertoires of many other primates. Some calls are tightly linked to a relatively narrow context, whereas others are used in a variety of circumstances. Some calls are directed at a specific individual, whereas others—including alarm calls— are more widely broadcast. Some calls are equally likely to be given by animals of all ranks and ages, others are not. Contact barks, for instance, are given by any individual who feels lost or separated. In contrast, threat-grunts are given only in aggressive interactions, and primarily by individuals who are higher-ranking than their opponents.

Unmodifiable, involuntary signals?

In marked contrast to children, who learn to produce and comprehend thousands of new words during their first three years of life, monkeys and apes rarely modify their vocal repertoires by adding new sounds. Although some primates make subtle modifications in their vocalizations as a result of experience (Hauser 1992a; Elowson and Snowdon 1994; Mitani and Brandt 1994; Crockford et al. 2004) and can modify the loudness of their calls through auditory feedback (Hage et al. 2006), a baboon in Kenya produces more or less the same sounds in the same contexts as a baboon in Botswana. This conclusion follows not only from research on many primate species (Seyfarth and Cheney 1997b) but also from a cross-fostering experiment involving two closely related species.

In this experiment, two infant Japanese macaques and two infant rhesus macaques were each cross-fostered into a group of the other species. The Japanese macaque infants were adopted by rhesus macaque mothers, and the rhesus macaque infants by Japanese macaque mothers. Japanese and rhesus macaques have very similar, baboon-like social structures. Each of the cross-fostered infants became fully integrated into its adoptive group. As they grew older, they even acquired ranks similar to those of their foster mothers. But while their social environment was similar to what they would have experienced in their own species' groups, the infants were exposed to a quite different culture of vocalizations. For example, although Japanese macaques give grunt-like gruffs in some contexts, they give a clear coo call when they play. Conversely, while rhesus macaques give coos in some contexts (particularly when feeding), they give gruffs when they play. What calls did the cross-fostered monkeys give?

Somewhat surprisingly, despite their ability to produce their adoptive species' calls, the cross-fostered juveniles continued to use their own species' vocalizations. In the rhesus macaque groups, the cross-fostered Japanese macaques gave coos while their rhesus playmates gave gruffs; in the Japanese macaque groups the opposite occurred. The cross-fostered animals behaved as if there were a rigid, unmodifiable link between call and context (Owren et al. 1992, 1993).

Monkeys, then, seem genetically predisposed to give particular calls in particular contexts. But this is not to say that their vocalizations are entirely reflexive and involuntary. Although their call *repertoire* may be relatively fixed, their choice of whether to call or remain silent is more flexible. Consider baboons' predator alarm calls—surely some of their most emotionally charged vocalizations (Fig. 45). Although baboons sometimes give seemingly uncontrollable, frenetic alarm calls to predators when they are under attack, such responses are by no means automatic. If the predator poses no immediate threat, then only a few,

Figure 45. A group of juvenile baboons gathers around a snake. Snakes often elicit no alarm calls from baboons. Photograph by Keena Seyfarth.

or sometimes no, individuals produce alarm calls. If baboons spot a predator at some distance, the few individuals who give alarm calls will do so in a sporadic, almost diffident manner. There is no obligatory link between the sight of a predator and the production of an alarm.

Stronger evidence for flexibility in the production of animal calls comes from experiments on the "audience effect" in birds and monkeys. Peter Marler and his colleagues presented male jungle fowl (*Gallus gallus*) with a silhouette of a hawk that "flew" over the birds' cage on a wire. The roosters gave alarm calls at high rates whenever they were in the presence of a member of their own species, but almost no alarm calls when they were alone (Gyger et al. 1986; Karakashian et al. 1988). When we carried out a similar experiment on an isolated pair of captive vervet monkeys, adult females gave more alarm calls to a simulated predator when they were with their own offspring than when they were with an unrelated juvenile (Cheney and Seyfarth 1985b). In both of these experiments, call production depended on the caller's audience. There was no obligatory, involuntary link between the sight of a predator and the production of an alarm.

Similarly, female baboons do not always give move grunts when initiating a move or infant grunts when approaching mothers with infants. Listeners seem to recognize that grunt production is variable and voluntary, and they respond accordingly. If a dominant female grunts to a lower-ranking mother as she approaches, the mother acts as if she recognizes that there is little likelihood that she will be threatened or have her food taken away from her. If, on the other hand, the dominant female remains silent, the subordinate takes this as a cue that she should leave, even if it means relinquishing a tasty fruit (Chapter 5). Far from being fixed and invariant, then, the production of a grunt depends on the details of the social situation and the particular individuals involved. Indeed, watching baboons grunt is rather like watching humans engaged in a conversation: animals grunt to one another in a calm, relaxed manner, and seem to grunt or remain silent entirely out of choice. Furthermore, baboons also seem to use grunts (and the lack thereof) to assess each other's intentions (Chapter 8).

In the absence of experiments, it is impossible to determine the extent to which monkeys have voluntary control over their vocalizations or whether they can explicitly decide whether or not to give a call. It remains entirely possible that female vervets give more alarm calls when they confront a predator in the presence of their offspring because they become more excited and afraid in this context than when they are with an unrelated juvenile. Similarly, female baboons may grunt to mothers with infants because grunts are difficult to repress when they are highly motivated to be friendly. In the laboratory, however, monkeys' vocalizations can be brought under operant control (Pierce 1985). It seems likely, therefore, that the spontaneous vocalizations of monkeys are also under some voluntary control.

This hypothesis is supported by a recent experiment in which two captive Japanese macaques were trained to use a rake to retrieve food. After training, the monkeys spontaneously began to give coos when using the rake. This was not surprising, because under natural conditions Japanese macaques give coos when foraging for and finding food. The experimenters then trained one of the monkeys to vocalize to request food or the rake. When they analyzed the acoustic features of the monkey's coos, they found that the monkey had, of his own accord, adopted one coo type to request food and another coo type to request the rake. Even though the monkey may have been strongly predisposed to give coos in the context of food, he seemed to have enough control over call production to use different coo types when making his requests (Hihara et al. 2003).

Further evidence that monkeys can choose when to call and when to remain silent comes from experiments performed by Wich and de Vries (2006) on groups of Thomas langurs (*Presbytis thomasi*), a species of Indonesian monkey that lives in small groups containing several females and one male. They exposed 12 different groups to a model of a tiger and found that, in each group, the male continued to give alarm calls until every other group member had also given at least one alarm. The males behaved as if they actively monitored the calling behavior of others and only ceased calling themselves after every other individual had acknowledged the predator's presence.

Emotional signals?

From at least Darwin's time, vocal communication in animals has been thought to differ from human language largely because the former is an affective system based on emotion, whereas the latter is a referential system based on the relation between words and the things they represent. Over the years, much ink has been spilled—by us and others—in debates about whether animal vocalizations could ever have a referential component and, if so, how referential and affective signaling might interact (see Seyfarth and Cheney 2003 for a review). Often the debate has been cast as an either/or dichotomy between affective and referential signaling. Such a dichotomy, however, is logically untenable.

A call's potential to function as a referential signal depends on the link between call type and social context. The mechanisms that underlie this link are irrelevant. A tone that informs a rat about the imminence of a shock, an alarm call that informs a vervet about the presence of a leopard, or a scream that informs a baboon that her offspring is involved in a fight all have the potential to provide a listener with precise information because of their predictable association with a narrow range of events. The widely different mechanisms that lead to this association have no effect on the signal's potential to inform.

Put slightly differently, there is no obligatory distinction between "referential" and "affective" signaling. Knowing that a call has the potential to convey highly specific information tells us nothing about whether its underlying causation is affective or not. Conversely, knowing that a call's production is due entirely to the caller's affect tells us nothing about the call's potential to function as a referential signal.

It is therefore wrong, on theoretical grounds, to treat animal signals as *either* referential *or* affective, because the two properties of a communicative event are logically distinct and independent dimensions. The first concerns the signal's relation to features of the environment, whereas the second concerns the underlying mechanisms by which that relation arises. Highly referential signals could, in principle, be caused entirely by a signaler's emotions, or their production could be relatively independent of arousal state. Highly affective signals could be elicited by very specific stimuli and thus function as referential calls, or they could be elicited by so many different stimuli that they provide listeners with only general information. In principle, any combination of results is possible. The "affective" and "referential" properties of signals are also logically distinct—at least in animals—because they may be different for signalers and listeners. The mechanisms that cause a signaler to vocalize do not in any way constrain a listener's ability to extract information from the call.

Once again, baboon grunts offer a good example. Drew Rendall (2003) used behavioral data to code a social interaction involving move or infant grunts as one of high or low arousal. He then examined the calls given in these two circumstances and found that in each context certain acoustic features or modes of delivery were correlated with apparent arousal. Bouts of grunting given when arousal was seemingly high were characterized by more calls, a higher rate of calling, and calls with a higher fundamental frequency (F0) than bouts given when arousal appeared to be low. Further analysis revealed significant variation between contexts in the same three acoustic features that varied within context. By all three measures (call number, call rate, and F0), infant grunts were correlated with higher arousal than were move grunts. Infant grunts also exhibited greater pitch modulation and more vocal "jitter" (Rendall 2003).

It is, of course, difficult to obtain independent measures of a caller's arousal in the field. However, similarities between human and non-human primates in both the mechanisms of phonation (Schön Ybarra 1995; Fitch and Hauser 1995; Fitch et al. 2002) and the expression of emotions (Scherer 1989; Bachorowski and Owren 1995; Hammerschmidt and Jurgens in press) support Rendall's (2003) conclusion that different levels of arousal play an important role in causing baboons to give acoustically different grunts in the infant and move contexts. But accepting this view says nothing about the grunts' potential to act as referential signals that inform listeners about events taking place at the time. Move grunts are linked to the context of group movement and therefore have the potential to convey quite specific information to listeners (Fig. 46). When Comet hears Balo give a move grunt, she learns with considerable precision what is happening in Balo's life at that moment. By comparison, infant grunts are not as tightly linked to a particular type of social interaction, and as a result their meaning is less precise. When Comet hears Balo's infant grunt, she learns that Balo

Figure 46. Baboons often give move grunts when traveling through wooded areas where they are out of sight of others. Photograph by Roman Wittig and Cathy Crockford.

is involved in some sort of friendly interaction, but the precise nature of the interaction is unknown.

"Discrete" versus "graded" signals

Historically, the dichotomous view that language is referential whereas animal vocalizations are emotional has gone hand in hand with the notion that human speech is perceived as a number of discretely different sounds whereas animal vocalizations are perceived as an acoustically graded continuum. But this dichotomy, too, is overly simplistic and largely incorrect. Just as a baboon's grunt may reflect her level of arousal and yet convey specific, referential meaning to a listener, so a baboon may produce a graded continuum of sounds that is nonetheless perceived by listeners in a discrete, categorical manner. This is not to say that perception of a graded continuum of sounds is *always* discrete, just that the production of an acoustically graded continuum does not in itself rule out the possibility of perception in terms of discrete categories (Marler 1976). In fact, such perception appears to be widespread (Fischer 2006).

Julia Fischer's (1998) study of Barbary macaque alarm calls was one of the first to demonstrate "categorical perception" of a graded series of sounds in free-ranging nonhuman primates. Since then, this result has been replicated several times, most notably in studies showing that baboon listeners respond to several of their acoustically graded calls as if they provide discretely different sorts of information. For example, although baboons' infant and move grunts grade acoustically into one another (Owren et al. 1997), listeners treat them as if they provide qualitatively different sorts of information. In playback experiments, listeners are significantly more likely to "answer" a move grunt than an infant grunt by giving a call of their own, and they are significantly more likely to scan the surrounding area after hearing a move grunt than after hearing an infant grunt (Rendall et al. 1999). At the same time, listeners' responses are also affected by context, because females are more likely to give answering grunts and scan the areas around them when the group is moving than when it is resting. Thus, while the baboons' move and infant grunts grade acoustically into one another, the two calls nonetheless function as discretely different signals whose meaning is influenced by the context in which they occur.

Similar results come from another playback study in which Julia Fischer divided female baboons' alarm and contact barks into those that unambiguously belonged to the contact or alarm category and those that were acoustically intermediate. She conducted playback experiments and found that infants responded in qualitatively different ways, and with increasing intensity, to typical contact barks, intermediate barks, and typical alarm barks (Fischer et al. 2000). Adults, on the other hand, showed strong responses only to alarm barks. Their responses to intermediate and contact barks depended upon the context. If an adult was alone on an open savannah and particularly vulnerable to predation, she treated an intermediate bark as if it were an alarm bark. If she was with other animals or foraging in a relatively safe place, she did not. These results suggest, again, that although alarm and contact barks show acoustic intergradation, listeners treat them as if they provide discretely different information (Fischer et al. 2001b). And, as with grunts, a call's meaning (as measured by listeners' responses) depends on a combination of information derived from the call itself and information derived from the context in which it occurs.

In a third study, Dawn Kitchen showed through playback experiments that the alarm and contest wahoos of adult males, which grade acoustically into one another (Fischer et al. 2002), are nonetheless perceived by female baboons as discretely different vocalizations that require qualitatively different responses (Kitchen et al. 2003a).

The acoustic intergradation in all of these vocalizations may be caused by gradation in the caller's arousal or emotional state. Indeed, Fischer's analysis suggests strongly that the acoustic variation in female contact and alarm barks, and in male contest and alarm wahoos, is consistent with an explanation based on variation in the caller's emotions (Fischer et al. 2001a, 2002, 2004a; see also Jurgens 1995). Rendall (2003) also argues convincingly that the acoustic variation in infant and move grunts is best explained by assuming that signalers giving infant grunts are more emotionally aroused than signalers giving move grunts, and that within the former category some infant grunts reflect more excitement than others. In each of these cases, however, graded calls whose production may be determined largely by the signaler's emotions are nonetheless perceived as discretely different vocalizations. And once listeners have recognized that each of these discretely different signals is predictably linked to a particular event, calls have the potential to convey to listeners a meaning that goes far beyond information about the signaler's emotional state.

There is, finally, yet another way in which listeners respond to a graded continuum of calls as if each call provides discretely different information. As we have mentioned, most if not all of the baboons' vocalizations are individually distinctive. When a baboon hears a call, she knows immediately that the caller is Sylvia and not Hannah, Fat Tony, or any other group member. She knows, in other words, that the call belongs in one of approximately 80 discretely different categories. Individual recognition by voice is a fact of life for baboons and other primates and, as we have seen, caller identity can strongly affect a call's meaning. After Hannah has fought with Nimi, Hannah's grunt tells Nimi that Hannah is unlikely to threaten her again, whereas a grunt from Beth is largely irrelevant, at least for the moment. In this respect as in many others, baboons have been selected to arrange a graded series of calls into discrete categories. The graded information is discretely coded because there are no intermediate values: the call that Nimi hears is either from Hannah or Beth, not some intermediate chimera.

Given the potential ambiguity inherent in a graded series of calls, and the importance of distinguishing both between different call types and between the calls of different individuals, it appears that baboon listeners have been under strong selective pressure to detect subtle distinctions within a graded acoustic continuum and to link these differences in acoustic structure with differences in individual identities, social events, predators, and so on. We are only beginning to understand how this linkage takes place. Neurophysiological research (discussed below) is beginning to reveal how the brain encodes call "meaning" in non-human primates, but we still know very little about how this encoding relates to our human conceptions of "emotional" and "referential" information. A baboon may recognize a grunt as Sylvia's because *any* vocalization by Sylvia elicits high anxiety. Alternatively, she may recognize Sylvia's grunt because, in a much less emotionally charged way, hearing it conjures up in her mind a vision of the aged curmudgeon. Referentiality in baboons and other primates could be achieved by the encoding of emotional information that is tightly linked to specific external events (recall Premack's argument about the strawberries), or it could be largely independent of emotion; at this stage, we simply do not know. In the following sections we take a closer look at the "meaning" of primate signals.

Meaning and emotion: The argument so far

To this point, we have argued that the dichotomy separating "involuntary" primate vocalizations from "voluntary" human speech is misleading. Primate vocalizations are not involuntary reflexes, impossible to suppress. They are, instead, much more like the other behaviors in which animals choose to engage. As they go about their daily lives, baboons decide whether or not to vocalize, just as they decide whether or not to groom, play, or form an alliance. Their behavior depends on a complex combination of their own motivation, the particular situation at hand, and who else is involved. Primates can control whether they vocalize or not; what they cannot control are the detailed acoustic features of the calls they choose to produce.

The dichotomy separating "graded" animal vocalizations from "discrete" human speech is equally misleading, because a graded continuum of calls can nonetheless be perceived in a discrete manner. Baboon listeners, for example, often respond to acoustically graded signals as if they convey discretely different information, and they assign unique, categorical, individual identities to different signalers' calls.

Finally, the dichotomy separating "emotional" primate calls from "referential" human speech is both logically and empirically false because, regardless of the mechanisms that underlie their production, vocalizations can provide listeners with highly specific information about events external to the signaler. Baboon vocalizations acquire their meaning because each call is individually distinctive and each call type is predictably associated with a specific social context. Listeners recognize these associations. They imbue calls with meaning.

The representation of meaning in animal vocalizations

Meaning, however, is not a term to be used lightly, especially in the company of philosophers and linguists. When Pavlov's dog salivated at the sound of a metronome, Pavlov did not rush to conclude that the dog interpreted the sound as a symbol for meat in the same way that the word *steak* is a symbol for meat to humans. Instead, he cautiously concluded that the dog had formed an association between the metronome and meat, with the result that the sound alone came to elicit the same response as the meat itself.

Thus far, the data we have reviewed are no different from Pavlov's. When a baboon hears an alarm bark and immediately runs to a tree, her response could simply be the result of a learned association between the sound and imminent danger. Nothing in the baboon's behavior forces us to conclude that she understands the sound-meaning relation that links a particular call with a specific feature of the world. What kind of experiment would allow us to test this hypothesis?

When we hear a word, we process it simultaneously at two levels. At the acoustic level we hear it, take note of how it sounds, and distinguish it from other words that have a similar sound. At a higher, more abstract level—call it the semantic level—we take note of what the word means. If you were asked whether the words *treachery* and *deceit* are similar or different, you would probably answer that they are alike. You would ignore the fact that they sound different and focus instead on the fact that they mean roughly the same thing. And if you were asked to compare *treachery* and *lechery* you would probably ignore the fact that they sound alike and say that they were different—unless, of course, you were writing a poem about a dissipated traitor and looking for words that rhyme.

When does a sound cease to be just a sound and become a word? David Premack (1976) argued that this transformation occurs when the sound is judged, or two sounds are compared, not on the basis of their physical properties (how they sound) but according to the properties of the objects they denote (what they mean). We can be fairly confident that *treachery, deceit,* and *lechery* have semantic meaning to humans because they can tell us that they regard *treachery* and *deceit,* but not *treachery* and *lechery,* as synonyms. Can monkeys make similar classifications?

Figure 47. A Diana monkey from the Tai forest, Ivory Coast. Photograph by Florian Moellers.

Klaus Zuberbuhler tested this hypothesis in a series of experiments on wild Diana monkeys (Fig. 47). To begin, he played from a concealed loudspeaker the growling of a leopard and found that it elicited a chorus of leopard alarm calls from nearby monkeys. Then he played the shriek of a crowned eagle and showed that it elicited a chorus of eagle alarm calls. Next, Zuberbuhler played a male Diana monkey's leopard alarm call and found that it elicited a chorus of leopard alarms from females. This was interesting because the female's leopard alarm call is acoustically different from the male's, so the females were not just imitating the male's call. Similarly, when females heard a male Diana monkey's eagle alarm call, they responded with their own eagle alarm call, which, again, is acoustically different from the male's.

Zuberbuhler then designed an experiment that, in effect, asked the *treachery-deceit* question. It asked the monkeys to compare two acoustically different calls that seemed to denote the same object (leopard growls and a male's leopard alarm; eagle shrieks and a male's eagle alarm), and two different calls that seemed to denote different objects (leopard growls and a male's eagle alarm; eagle shrieks and a male's leopard alarm). His method took advantage of the fact that Diana monkeys, like most animals, habituate to the same stimulus if it is presented repeatedly. Diana monkeys respond strongly when they first hear a leopard's growl. But if they hear the same call several minutes later they remain silent, presumably because the second call provides them with no new information.

When females heard a male's leopard alarm, they responded by giving their own leopard alarm calls. Five minutes later, however, when the females heard a leopard's growl coming from the same area, they gave no response. The females remained silent even though they would have given leopard alarm calls if they had heard the growl in the absence of any prior alarm calls. In contrast, when the females heard an eagle's shriek five minutes after hearing a male's leopard alarm, they responded with a chorus of their own eagle alarm calls. Similarly, females who had first heard a male's eagle alarm call did not respond, five minutes later, to the shriek of an eagle. But they did respond with leopard alarm calls if they were played the growl of a leopard. Diana monkey leopard alarm calls and leopard growls sound very different, as do Diana monkey eagle alarm calls and eagle shrieks. Nonetheless, the monkeys responded to the two leopard-associated noises and the two eagle-associated noises as if they provided the same information—as if they were synonyms. They classified sounds on the basis of their meaning, not just their acoustic properties (Zuberbuhler et al. 1997, 1999).

Zuberbuhler's results support a *representational* theory: a theory in which organisms have mental representations of objects in the world and these representations influence their behavior. The theory argues that when a female Diana monkey hears a sound like a male's leopard alarm call, two things happen: she recognizes the link between this sound and an actual leopard, and she represents a leopard in her mind. To paraphrase the philosopher Jerry Fodor, the female Diana monkey introduces a *semantic connection* into the causal chain that begins when she hears the alarm call. The connection is semantic because in this chain of events the leopardness of the sound appears twice, once by virtue of the way things are in the world and once as it is represented in the monkey's mind (Fodor 1986). The mental representation has causal power because when the female hears a leopard's growl coming from the same location five minutes later, she compares this new representation with the one she already has. Since the new information is redundant, the female does not respond to the growl by giving an alarm call. Instead, she remains silent. Had the growl been an eagle's shriek, however, the female would have responded by giving her own eagle alarm call.

Notice that in Zuberbuhler's experiment there is no fixed, invariant link between stimulus (a leopard's growl, for instance) and response (an alarm call of a specific sort). If a monkey hears a growl, she may give an alarm call, or not. We can explain the monkey's selective responding only if we assume that it is caused by an intervening psychological process that depends crucially on stored information about sounds and what they stand for. Stated differently, we can explain the monkey's behavior only if we assume that, in the mind of a listener, calls have acquired *meaning*.

This description of the information acquired during communication does not require any specialized skills on the part of monkeys. To the contrary, it is entirely consistent with modern versions of Pavlovian theory as it has been developed on rats and pigeons (Dickinson 1980; Rescorla 1988). If a rat learns that a light signals the imminent delivery of a shock, the rat begins to show fear and avoidance whenever the light comes on; the light provides the rat with information about the shock. Further, if a rat first learns that a tone is associated with the light and then learns that the light is associated with shock, it exhibits fear and avoidance behavior not only when the light comes on but also when the tone is heard, even though tone and shock have never been associated directly.[1] Although the light and the tone are qualitatively different sorts of stimuli and are processed by different sensory mechanisms,

---

1. See Rizley and Rescorla (1972). Brogden (1939) was the first to describe this phenomenon. He called it "sensory preconditioning." Note that the experiments on rats and pigeons do not rely on habituation. They therefore argue against the view that, in studies like Zuberbuhler's, Diana monkeys remain silent after a hearing a leopard's growl simply because they just gave a leopard alarm a few minutes ago and animals get bored producing the same call over and over (Thompson 1995; Heyes 1994; Schusterman and Kastak 1998).

they have come to elicit the same response because they provide the rat with the same information.

Criticism of the representational theory

Our application of the representational theory to monkeys and other animals can be criticized on several grounds. To begin with, it is vague: although we talk about the "meaning" of vocalizations to monkeys, we cannot say precisely what they mean or specify exactly where in the brain such mental representations reside. Our discussion of call meaning can also be criticized as simplistic and premature, because we have borrowed a complex term from linguistics and philosophy without pointing out the substantial differences between humans' use of words and animals' use of vocalizations. Below we consider each of these issues in turn. To preview our argument, we accept the notion that our use of "meaning" is imprecise (though this also holds for words). At present, scientists know much more about the neural basis of meaning in the human brain than about its counterpart in the brains of monkeys, but preliminary data reveal some striking similarities. Finally, we borrow the representational theory to highlight not just the similarities and differences between baboons and humans but also to bolster the argument that baboons' vocalizations provide the key that unlocks their minds.

What information, exactly?

In *Word and Object* (1960), the philosopher W. V. O. Quine imagines a linguist who has arrived in a foreign land where people speak a completely unknown language. The linguist sets out to discover what their words mean. But when a rabbit runs by and a local citizen shouts "Gavagai!" the linguist realizes that determining precisely what this word means is going to be more difficult than he had anticipated. *Gavagai* might mean *large ears, see how fast it runs, hippity-hop, let's catch it and eat it,* or even *rabbit.* In principle, an infinite number of meanings is possible. Quine's point was that, even with an intimate knowledge of the language, the linguist could never be certain that his meaning of *gavagai* was exactly the same as another's. The linguist might be able to specify the conditions under which *gavagai* was uttered, but his assessment of the word's meaning would always be, in comparison with other relations in the natural sciences, an approximation. Quine called this the "radical indeterminacy of meaning."[2]

Other philosophers, like Ludwig Wittgenstein (1953) and George Herbert Mead (1964), took a more social, contextual approach to the study of meaning. They argued that the meaning of a word depends not only on its link to a specific referent but also on its use—in a sentence, by a particular speaker, in a particular social context. These contextual variables lead inevitably to slight modifications in meaning from one communicative event to another: the same word never means exactly the same thing in two successive uses. Unlike Quine, Wittgenstein and Mead were untroubled by this approximation because, regardless of how nuanced a word's meaning might be, speakers and listeners must still agree sufficiently for the word to function in natural conversation and be passed down from generation to generation (Pears 1987). Radically indeterminate they may be, but words still *work*.

Given the difficulty of specifying the meaning of human words, any attempt to specify the meaning of animal vocalizations is bound to be even more problematical, because we cannot hope to understand another species' mind as well as we understand our own. Making matters worse, while some animal vocalizations (like predator-specific alarm calls) are given in narrowly defined circumstances and appear to have a fairly specific meaning, many others (like baboon infant grunts) are only loosely linked to particular events. The radical indeterminacy of many animal vocalizations should not, however, deter us from studying their meaning, for at least three reasons.

First, as Quine pointed out, radical indeterminacy is everywhere, from the normal speech of adults to the one-word utterances of young children and the confused ramblings of neurological patients. This makes the study of meaning difficult but it does not make it impossible, nor does it rule out such research in principle.

Second, we should not underestimate the rich information that vocalizations can provide to listeners even in cases where a call's meaning is imprecise. As we have mentioned, many animal vocalizations, includ-

---

2. Sixty-eight years earlier, in *The Speech of Monkeys* (1892), Garner anticipated Quine's example: "If you should be cast away upon an island inhabited by some strange race of people whose speech was so unlike your own that you could not understand a single word of it, you would watch the actions of those people, and see what they did in connection with any sound they made; and in this way you would gradually learn to associate a certain sound with a certain act, until at last you would be able to understand the sound without seeing the act at all." Like others, Garner was untroubled by the radical indeterminacy of meaning.

ing those of baboons, are individually distinctive. Beginning with this observation, scientists studying birds, primates, and other animals have conducted playback experiments to test hypotheses about individuals' knowledge of their social companions. By using combinations of calls designed to provide listeners with the information that a certain event has taken place, experiments have shown, for example, that birds eavesdrop on the singing contests of rivals to assess their relative dominance ranks (Chapter 7), vervet monkeys remember which individuals have recently groomed them (Chapter 6), and baboons recognize the rank and kin relations that exist among others (Chapter 6). All of these studies presume that listeners gain at least one bit of very precise information when they hear a call: information about the identity of the signaler. From this relatively simple starting point they go on to derive much more complex information about the social events taking place around them, as well as the social relationships that these events imply.

A third reason for studying the meaning of animal calls concerns the future direction of comparative neurobiological research. If we accept the view that meaning plays no role in animal communication, it follows that the semantic information conveyed by human words is unique, with no parallels in the vocalizations of any nonhuman creatures (for examples of this argument, see Owings and Morton 1998; Owren 2000). This conclusion, though, is premature, as we shall see in the following section.

What, precisely, are "mental representations"?

It is time to admit that, despite the ubiquitous use of the term, no one can say precisely what mental representations actually are, or where they are stored in the brain. But this is not necessarily a problem. There is, in fact, a long history in science of giving a name to something without really knowing what it is. In 1909, the Danish biologist Wilhelm Johannsen christened Mendel's units of heredity "genes." He had no idea what genes were; all he knew was that they obeyed certain laws. Over the next 40 years, "like the atoms of the physical chemist at much the same time, genes acquired combinatorial rules and otherwise became imaginable as their manifestations were observed" (Judson 1979:206). There is, then, nothing wrong with using a term like "mental representation" to label an entity whose physical properties we are only beginning to understand.

Research on the brain structures that underlie monkeys' knowledge of the world has lagged behind research on the human brain, for several reasons. First, many of the most exciting developments in human brain function have come from the application of functional magnetic resonance imaging (fMRI). This technique has only recently been applied to monkeys (Logothetis et al. 2001; Ghazanfar and Logothetis 2003). Second, human language, including our enormous vocabulary, allows scientists to present subjects with extremely specific questions about familiar objects or problems while they measure brain activity. Comparable experiments on monkeys are difficult to conduct—difficult, but not impossible, as we shall see below.

From fMRI studies of the human brain, there is growing evidence that we encode object words using a distributed neural representation that combines "a perceptual representation based on an object's physical features and a semantic representation based on previously acquired information about the object" (Martin 1998:72; see also Barsalou et al. 2003; Caramazza and Mahon 2006; Humphreys and Riddoch 2006). Perceptual information about the object's physical features is distributed among a number of different brain areas depending on the specific feature involved. Information about the object's color is stored close to the brain region that mediates color perception, information about how the object moves is stored close to the brain region that mediates perception of motion, and so on. The attributes and features that uniquely define an object, distinguishing it from other objects in the same general category, are bound together as a result of having been experienced together when the object was previously perceived. This binding creates "a network of discrete cortical regions: a semantic network" (Martin 1998:74).

The human semantic network is activated automatically, not only when an object is seen but also whenever its name is read, heard, or retrieved as we speak or write. The network, in other words, is multimodal. Hearing the sound of a train causes activity in both brain regions associated with audition and brain regions associated with visual perception and the perception of motion. Seeing a train (or a picture of a train) produces activity in visual areas and also makes it easier for us to imagine the train's chugging and whistling sounds (Barsalou et al. 2003).

The human semantic network also appears to involve automatic, unconscious activation of previously acquired information about an object. In experiments that asked human subjects to identify line drawings that they had never seen before, subjects took less than 0.7 seconds to identify what the drawings represented. "How," the neuroscientist Alex Martin asks, "could we quickly recognize or identify an object as being of a particular type (chair, pencil, kangaroo, etc.)" if our semantic network did not instantly provide us with information derived from our prior knowledge of chairs, pencils, and kangaroos (Martin 1998:74)?

Although research on the representation of call meaning in monkeys has only just begun, some intriguing parallels between monkey and human processing have already emerged. Like humans, rhesus macaques have areas in the brain that respond selectively to their own species' vocalizations (as compared with other auditory stimuli) and to their own species' faces (as compared with other visual stimuli) (see Chapters 1, 7, and 8). Different sorts of face cells in different areas of the rhesus macaque temporal cortex respond selectively depending upon facial identity, facial expressions, and gaze direction—strong indications that the rhesus' brain has evolved specific structures to deal with particular aspects of its social world (see Ghazanfar and Santos 2004 for review). Like humans, monkeys exhibit neural lateralization (a left-brain, right-ear advantage) in the processing of their own species' calls but not in the processing of other auditory stimuli (Petersen et al. 1978; Hefner and Hefner 1984; Weiss et al. 2002; Poremba et al. 2004).

Perhaps most important, rhesus macaques appear to process and categorize sounds in ways that are analogous to humans' processing of speech sounds. When rhesus macaques hear their own species' calls, they show a pattern of neural activation that is strikingly similar to the one described for humans. As we described in Chapter 8, the monkeys show neural activity not only in areas associated with auditory processing but also in areas necessary for visual processing, and for storing information about objects in memory (Gil da Costa et al. 2004). And anatomical studies of the brain reveal strong connections between these different areas (Poremba et al. 2003). The result is a multimodal pattern of activation that integrates auditory and visual information (Ghazanfar and Logothetis 2003; Ghazanfar et al. 2005; Barsalou 2005; see also Barraclough et al. 2005).

Furthermore, when rhesus macaques hear their own species' calls they appear to evaluate them according to their "meaning" rather than their acoustic features alone. Like Diana monkeys, rhesus macaques discriminate between different calls on the basis of the information they convey (Gifford et al. 2003). Yale Cohen and his colleagues (Gifford et al. 2005; Cohen et al. 2006) played food-associated calls and calls given in nonfeeding contexts to rhesus macaques and examined the resulting activity in the monkeys' ventrolateral prefrontal cortex (vPFC), an area of the brain associated with the classification of stimuli (Freedman et al. 2001). The food-associated calls included coos and gruffs, two call types that are acoustically very different. They found that neurons in the vPFC responded differently to food-associated calls than to other call types despite their acoustic differences, indicating that the prefrontal cortex plays an important role in the classification of auditory stimuli by function, or "meaning." These results are particularly interesting because the monkeys were not trained to make these classifications; they did so spontaneously.

Taken together, these studies support the suggestion that monkeys do more than just perceive a sound at the acoustic level when they hear a particular call type. They also form a multimodal (visual and auditory) representation of what the sound means. There is, therefore, considerable support for the existence of "a homologous system in non-human primates and humans for representing object information" (Gil da Costa et al. 2004:17518). Or, as Barsalou (2005:309) suggests, when one monkey hears another vocalizing "the auditory system processes the call, the visual system processes the faces and bodies of conspecifics, along with their expressions and actions, and the affective system processes emotional responses. Association areas capture these activations as they occur repeatedly, storing them for later representational use. When subsequent calls are encoded, they reactivate the auditory component, ... which in turn activates the remaining component in other modalities. Thus the distributed property circuit that processed the original situation later represents it conceptually."

In sum, while we do not yet know precisely what mental representations are, modern neurobiology, greatly aided by new developments in brain imaging, is beginning to reveal the structure of mental representations in humans and in monkeys. But are monkey vocalizations really like words? If the same cortical network were activated in a baboon's mind when she saw a lion and when she heard a baboon's alarm bark, would we be right to conclude that the alarm bark meant *lion* to the baboon in the same way that the word *hippo* denotes a particularly malevolent herbivore to us? Although the parallels in brain function would be compelling, the answer must surely be "no." Why?

Meaning in linguistic and nonlinguistic creatures

To answer this question, we return to the observation that "meaning" is a loaded term in linguistics—encompassing far more than just information about objects or events in the world (for a valuable discussion of these issues, see Jackendoff 1994 and Pinker 1994). When John tells

Theo "The Red Sox relievers were lousy last night," he is not just mapping words onto events in the world. Much more is communicated than these bare facts. John's words are motivated by his desire to provide Theo with information both about certain events and about his attitude toward them. From John's statement, Theo learns not just about the sorry state of the Red Sox bullpen but also about John's unhappiness with the Red Sox roster. John assumes that Theo hasn't yet heard about the Red Sox's late inning swoon. Theo assumes that John is disgruntled because he is a Red Sox fan.

Mind-reading pervades language. Because humans rarely speak or listen to others without attributing mental states to them, the meaning of words in language includes far more than just the mapping of a sound onto a referent. Indeed, the philosopher H. P. Grice (1957) argued that true linguistic communication cannot occur unless both the speaker and the listener take into account each other's state of mind. According to this view, monkeys and other animals cannot be said to communicate like humans unless they use calls with the intent to provide (or manipulate) information. Given that monkeys lack a full-blown theory of mind and cannot distinguish between their own knowledge and somebody else's (Chapter 8), even calls that serve a referential function may be based on mental mechanisms that differ fundamentally from those found in human speech.

There is another sense in which human words differ fundamentally from animal vocalizations. Words are more than just labels for concepts; they acquire additional meaning through their relation to other words and their functional roles as nouns, verbs, and modifiers (Pinker 1994). Words in a sentence play two roles simultaneously: they designate features of the world and they specify the roles these features play in a narrative event. Sentences have a complex, hierarchical structure that maps onto semantic structure. As a result, they are infinitely more powerful communicative devices than single words because they allow us to describe not just things but relations among things. Sentences allow us to take a relatively finite vocabulary and generate an infinite number of messages.

But to speak and understand sentences, individuals must understand the rules, or grammar, that underlies them. Creating and decoding sentences, moreover, is complicated because the words in a sentence are not simply joined together like links in a chain. Instead, they are grouped into phrases and organized into a nested, hierarchical structure. Linguists depict this phrase structure by means of a tree diagram, so called because of its branching appearance. For example:

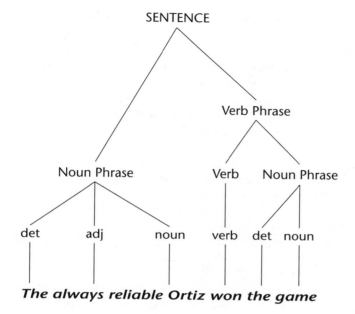

SENTENCE

Verb Phrase

Noun Phrase   Verb   Noun Phrase

det   adj   noun   verb   det noun

The always reliable Ortiz won the game

Underlying phrase structure are rules governing the use of its components. A noun phrase consists of an optional determiner (det), followed
by any number of adjectives, followed by a noun. A verb phrase consists
of a verb followed by a noun phrase. A sentence consists of a noun
phrase followed by a verb phrase. For present purposes—interested, as
we are, primarily in baboons—three features of this phrase structure are
important. First, as the linguist Steven Pinker points out, the elements
that make up the tree are modular. "A symbol like NP (noun phrase) is
like a connector or fitting of a certain shape. It allows one component
(a phrase) to snap into any of several positions inside other components (large phrases). Once a kind of phrase is defined by a rule and
given its connector symbol, it never has to be defined again; the phrase
can be plugged in anywhere there is a corresponding socket" (Pinker
1994:98–100). A noun phrase is a noun phrase—and obeys the rules
that govern noun phrases—regardless of whether it comes at the start of
the sentence (the actor) or embedded in a verb phrase (the acted upon).

Second, tree structure diagrams demonstrate clearly how words derive their meaning both from their link to objects in the world and
from the role they play in a particular grammatical structure. In the
sentence "Two cars were reported stolen by the police," the phrase "by
the police" could be placed in either of two locations in the tree struc-

ture: as part of the verb phrase beginning with "reported" or as part of the verb phrase beginning with "stolen." Where it goes affects the sentence's meaning (Pinker 1994:102). Furthermore, the component of a word's meaning that derives from its grammatical role can be separated entirely from the component of its meaning that derives from its link to the outside world. Chomsky's famous sentence "Colorless green ideas sleep furiously" makes grammatical sense and conveys meaning—it is the ideas that are sleeping furiously—even though the events it describes bear no relation to reality.

Third, the cognitive abilities that allow us to construct and decode sentences are, to a great extent, innate properties of the human mind. Many different sorts of data converge to support this conclusion. Children learn their first language effortlessly and without explicit awareness of the grammatical rules that underlie it. Indeed, long before they can speak grammatical sentences, children can distinguish grammatical from ungrammatical sentences spoken by adults. Language development unfolds surprisingly quickly, in the second and third years of life, following roughly the same order and timing regardless of the specific language learned or the degree of parental involvement. Children born blind and children born deaf who learn a sign language like ASL show a pattern of language development that may be slightly delayed but is similar in all important respects to language development in sighted and hearing children. And children exposed at an early age to a system of communication that lacks the formal properties of language—like a pidgin language, whether spoken or signed—spontaneously convert it into a full-blown language without explicit instruction. Finally, there are several sorts of genetic and neurological impairments that damage a person's language but leave his other cognitive abilities intact. Others damage some aspect of cognition but have little effect on language (Caplan 1992).

We have, it appears, a largely innate ability to take a string of sounds (or a string of hand gestures, in the case of deaf people) and do two things with astonishing speed: recognize the link between some sounds and features of the world; and organize the sounds into phrases that determine how they function in the sentence and what the sentence means. For the most part, we perform these tasks both rapidly and unconsciously, without being aware of what we are doing or how we do it (Pinker 1994; Jackendoff 1994; for another view see Tomasello 2003, 2005).

The relatively innate, specialized sentence-processing mechanisms found in human language raise two questions regarding baboon metaphysics. First, are there any things like sentences in the communication systems of baboons, other primates, or indeed any nonhuman creatures? The answer, as we will see, is essentially "no." Second, if our ability to parse sentences is unique in the animal kingdom, does this mean that it evolved only recently, after the ancestral line leading to modern humans had diverged from the common ancestor of human and chimpanzee? Or are there circumstances in the lives of baboons or other primates in which we can see the cognitive precursors of grammar? We discuss these questions further in the next chapter, where we consider the evolution of language.

ELEVEN

Precursors to Language

Plato says in Phaedo that our "necessary ideas" arise from the preexistence of the soul, are not derivable from experience—read monkeys for preexistence.
CHARLES DARWIN, 1838: *NOTEBOOK M*

In the mid-19th century, while Darwin and his contemporaries were arguing about Locke's and Kant's views of human metaphysics (see Chapter 1), an even more contentious debate raged on the origin of language. Strong views were expressed and scholars lost their composure. So acrimonious were the "discussions" that in 1866 the Société Linguistique de Paris banned all speculation on the topic. The decision held, and had long-lasting international repercussions. According to one rumor, the Linguistic Society of America considered a similar ban when it was founded in 1924, but settled instead for a "gentleman's agreement" prohibiting papers devoted to the origins of language. No paper on language evolution appeared in the society's journal *Language* until the year 2000 (Newmeyer 2003).

Darwin believed that the course of evolution could be revealed through the comparative method—by contrasting similar traits in related species, examining their common properties, measuring their differences, and searching for branching points in the fossil record. When the trait in question is language, the Darwinian approach might logically begin with a comparative study of human language and nonhuman primate vocalizations. But this technique has not proved very successful. The two sorts of communication are so different that comparison between them reveals little about their common ancestry. There is, however, another, more indirect, approach—one that examines the evolution of language through its links with cognition.

In humans, language and cognition are inextricably bound together. Through language, we express our thoughts and reveal how we see the world. Language thus offers a window into the mind, a view of mental content. When we use language we perform some of the most complicated mental gymnastics of which humans are capable. Language therefore tells us something about how the mind works. Baboons' cognitive skills may be much more elaborate and flexible than their vocalizations, but we can still use their communication to understand the content of their knowledge and how their minds work. This, after all, is the rationale behind playback experiments.

The close links in both humans and baboons between communication and cognition raise the possibility that we can apply the comparative method not by contrasting baboon vocalizations directly with language but instead by comparing the knowledge and thought that underlie baboon communication with some of the cognitive operations that are central to human language.

It is now well accepted that before a child can learn language she must have some experience with the objects, events, and relations that make up her world. As several linguists have put it, "If you couldn't pick pieces of meaning out of the world in advance, before you learned a language, then language couldn't be learnt" (Fisher and Gleitman 2002:447, quoting Chomsky 1982:119). The same argument appears in theories of language evolution. Pinker and Bloom (1990:713), for example, propose that "grammar exploited mechanisms originally used for ... conceptualization," and Newmeyer (1991:10) states that "the conditions for the subsequent development of language ... were set by the evolution of ... conceptual structure. A first step toward the evolution of this system ... was undoubtedly the linking up of individual bits of conceptual structure to individual vocalizations" (for similar views, see Jackendoff 1987, 2002; Kirby 1998; Newmeyer 2003; Hurford 1998, 2003).

The hypotheses that a certain kind of thinking appears before the emergence of language in young children, and evolved before spoken language in our hominid ancestors, are part of a broader intellectual tradition that can be traced to Plato, who believed that one could not acquire a concept that one could not antecedently entertain. More recently, this view has been articulated in the philosopher Jerry Fodor's influential book, *The Language of Thought* (1975). Fodor proposed that human knowledge and reasoning are couched in a "language of thought" that is distinct from external languages like English or Tsetswana. It contains symbols that pertain to people, objects, events, and the categories to which they belong, and causal relations that govern how objects behave in space and time and how people interact with one another.

The language of thought is distinguished from natural languages because of at least two prominent cases in which thought occurs without language. First, animals think, yet they have no language. Once we reject the behaviorist view that mental activities are an "explanatory fiction" and accept the proposition that animals possess information about each other, objects, events, and the relations between them, we commit ourselves to the existence of cognitive processes that are mediated by representational systems other than natural languages (Fodor 1975:57). Second, it is now widely accepted that before they learn language, human infants possess "core knowledge" that enormously aids their first categorization of objects, properties, and events in the world (Carey and Spelke 1996). As with animals, the central question then becomes how richly specified this private language (or "mentalese") may be, and how experience refines, enhances, and transforms it (Gleitman and Papafragou 2005).

At the same time, the language of thought is called a *language* of thought because it shares many properties with natural languages. Like natural languages, it includes some mental representations that correspond to objects in the world and others that specify the relation between these objects. Applied to humans, the language of thought hypothesis assumes that people—even preverbal infants—have "propositional attitudes": mental states with representational content. To have a certain propositional attitude is to be in a specified relation to an internal representation: to know, believe, or fear that something is the case (Fodor 1975; Stalnaker 1999). Both in the language of thought and in language, propositional attitudes are encoded in the form "thinks that...," "wants that...," and so on. The language of thought hypothesis helps explain why children learn their first language so easily (in some respects, they already have one), and why languages that differ in so many other ways are nonetheless similar in their basic function (Chomsky 1975).

Applied to theories of language evolution, the language of thought hypothesis predicts that knowledge of objects, events, and conceptual and causal relations preceded language, and that language evolved later as a means to express this knowledge (e.g., Jackendoff 2002:238). Obviously, we cannot test this view using modern humans, but we can approach the problem indirectly, as Darwin originally suggested, through the study of baboon metaphysics. That is what we propose to do in this chapter.

But first let us be clear about our methods. Because we are testing the language of thought hypothesis using modern-day baboons as our subjects, it might at first appear that we are treating them as living replicas of the 30-million-year-old common ancestor of monkeys, apes, and humans. Clearly, this would be a mistake. Baboons, after all, have undergone their own evolutionary changes in the years since they and we diverged from our common ancestor. Fortunately, though, reconstructing the common ancestor is not our primary goal. Instead, we study baboons in order to learn how evolution acts on the communication and cognition of animals that live in large social groups. If we can find general rules that specify how, in baboons, social complexity affects cognition and cognition affects reproductive success, then these rules might usefully be applied to research on the evolution of communication and cognition in many other species, including our own.

As we did in Chapter 10, in the sections below we draw some explicit comparisons between language and cognition in humans and the communication and cognition of baboons. We hope it will be clear that we are *not* arguing that baboons have a language, or that they are capable of anything close to linguistic communication—indeed, we devote as much time to the fundamental differences between baboons and humans as to their similarities. Rather than trying to claim that baboon communication is a lot like language, we propose the following argument:

1. Baboon vocal communication—and, by extension, that of other primates—is very different from human language. The differences are most pronounced in call production.
2. Differences in production have been overemphasized, however, and have distracted attention from the information that primates acquire when they hear vocalizations. In perception and cognition, continuities with language are more apparent.
3. In primate groups, natural selection has favored individuals who can form mental representations of other individuals, their relationships, and their motives.
4. This social knowledge constitutes a discrete, combinatorial system of representations—a language of thought—that shares several features with human language.
5. The language of thought that has evolved in baboons and other primates is a general primate characteristic whose appearance predates the evolution of spoken language in our hominid ancestors.

6. The prior evolution of social cognition created individuals who were preadapted to develop language.
7. Several features thought to be unique to language—for example, discrete combinatorics and the encoding of propositional information—were not introduced by language. They arose, instead, because understanding social life and predicting others' behavior requires a particular style of thinking.

Our focus on the early, prelinguistic period of language evolution means that we will not be discussing many features of modern languages whose evolution is hotly debated—features like case, tense, subject-verb agreement, open- and closed-class items, recursion, long-distance dependency, subordinate clauses, the subjunctive, subjacency, and the empty category principle. Nothing like these grammatical constructions exists in the communication of any nonhuman primate, including chimpanzees. They undoubtedly emerged during the later stages of language evolution, long after the divergence of the human lineage from the common ancestor of humans and chimpanzees.

So, rather than starting with modern language in all of its complexity and working backward—or, as Jackendoff (2002) puts it, searching within language for traces of its past—we begin instead with a system of communication that is clearly not a language but nonetheless shares some of its features. If we can show that at least some of the properties found in language can also be found in the social intelligence of nonhuman primates, we may gain a better idea of the cognitive foundations on which language was built.

Finally, some linguists believe that there is a fundamental problem in the application of evolutionary theory to language. For instance, Jackendoff (1990:737) states, "All the characteristics of organisms that have been examined for evidence of natural selection have been either physical structures or patterns of behavior. ... Linguistic theory, however, is not [just] about behavior, but rather about the mental representations that help determine perception and behavior." Here Jackendoff seems to be arguing that language is not so much a system of communication, on which selection pressures might have acted, as it is a system for mental representation and thought, where the role of evolutionary pressures is more difficult to imagine (Knight et al. 2000; see also Jackendoff 2002).

There is no doubt that language functions as both a means of communication and a tool for organizing and manipulating thoughts. But this in itself does not make it different from many other complex behaviors, nor does it make language immune to evolutionary explanations. Whenever an animal can gain a selective advantage by performing a particular behavior, natural selection will simultaneously favor both the behavior itself and whatever morphology, physiology, or cognitive operations are needed to support it. If nutcrackers can gain an evolutionary advantage by storing and recovering seeds from a huge number of locations, selection will simultaneously favor seed-hiding behavior, memory, and the neurophysiology to support it. And if the ability to represent, organize, and classify other individuals' relationships enables baboons to perform the mental calculations necessary for negotiating their social world, natural selection will favor whatever skills in communication, cognition, and neural machinery allow them to do so.

Call production

Repertoire size, development, and learning

Like other monkeys and apes, baboons make use of a relatively small number of calls. Depending on how one classifies call subtypes within a graded continuum (for example, move and infant grunts), a reasonable guess would be around 14 different vocalizations. In contrast, human speakers with an average vocabulary normally use about 50,000 words.

Primate vocalizations also differ from human speech in their development and in their flexibility during adulthood. It is well known that human children can learn to produce any of the thousands of phonemes used by *Homo sapiens* and thus can learn to speak any of the world's languages. Although human children are born predisposed to learn language (Jackendoff 1994; Pinker 1994; Bloom 2004), the particular language they learn depends crucially upon experience: they learn the language they hear. And language learning can be severely disrupted by abnormalities in social development (Curtiss 1977). Finally, once a child has learned his first language, he continues to add new words to his vocabulary throughout his life.

By contrast, vocal development in nonhuman primates seems relatively unaffected by variations in auditory experience or rearing (Chapter 10). This is not to say that the production of primate calls is entirely reflexive. Instead, as we discussed in Chapter 10, primate call production is both modifiable and innate. In any given context, individuals can choose to call or remain silent, but if they do produce a call its acoustic features are entirely predictable. Primates have a small repertoire of calls, and each call type is tightly linked to a particular social context. In these respects, primate vocal production is far less flexible and innovative than human language.

Words and sentences

Other differences between language and nonhuman primate communication are equally fundamental. Unlike human words, which are formed by combining smaller units (morphemes and phonemes) that typically have no meaning of their own, most primate calls are holistic. They cannot be broken down into their component parts, the way we can break down the word *walked* into the verb root *walk* and the past tense ending —*ed*. As a result, primates' calls are their communication system's smallest meaningful units.

Primates' calls are also, for the most part, their system's largest meaningful units. Unlike human speakers, who routinely combine words into sentences according to grammatical rules that allow them to describe relations among events, monkeys and apes almost never string different call types together in any rule-governed, structured way. A female baboon may give a rapid succession of grunts as she approaches a mother with infant, but she is not combining grunts in any linguistic sense. She is just saying "Baby, baby, baby, baby, baby, baby" or "want, want, want" over and over again. The number of repetitions may inform the mother that the female is intending to be extremely friendly rather than merely condescending, but for the most part several grunts convey the same message as one. No new meaning arises as a result of their combination.

One of the few exceptions to this generalization comes from the alarm calls of Campbell's monkey (*Cercopithecus campbelli*), a species that lives alongside Diana monkeys in the Tai forest of the Ivory Coast. Like the male Diana monkeys described in Chapter 10, male Campbell's monkeys have two acoustically distinct alarm calls, one given to leopards and another to crowned eagles (Zuberbuhler 2001). The alarm calls also elicit qualitatively different responses, both from other Campbell's monkeys and from Diana monkeys, with whom the Campbell's monkeys often associate. But there is a further, intriguing wrinkle to this story. When Campbell's monkeys are faced with a stimulus that is slightly alarming but not as serious as an actual predator—for instance, a falling tree, a breaking branch, the far-off alarm calls of another group, or the sound of a distant predator—they give a loud *boom!* vocalization, wait about 30 seconds, and then give one of their alarm calls. These preceding booms seem to modify the meaning of the subsequent alarm call because the same monkeys who respond strongly to a Campbell's alarm call when it is given on its own show little response to the same alarm call when it is preceded by a boom (Zuberbuhler 2002). The sequence "boom + alarm call" has a meaning that is different from the meaning of either a boom or an alarm call on its own. The call combination, in other words, carries a meaning that is more than the sum of its parts (Zuberbuhler 2003, 2005; for another example see Arnold and Zuber-buhler 2006).

Similarly, chimpanzees sometimes supplement their calls by drum-ming their hands and feet against resonant tree buttresses (Mitani 1993; Arcadi et al. 2004). In the Ivory Coast, male chimpanzees produce three acoustically different subtypes of barks: one when hunting, one when they encounter snakes, and a third, more generic bark in a vari-ety of different contexts. In two very limited contexts, when traveling or encountering a neighboring group, the chimpanzees combine their generic bark with drumming (Fig. 48) (Crockford and Boesch 2003). This signal combination may convey information that is qualitatively different from (and more specific than) the information conveyed by either the bark or drumming alone, although this hypothesis has not yet been tested experimentally.

Figure 48. A chimpanzee barks while drumming on a tree. Photograph by Roman Wittig and Cathy Crockford.

With these exceptions, however, primate speakers do not produce rule-governed call combinations. As a result, they have no means of marking different calls to designate their function in a call sequence as nouns or verbs, or as actors, actions, and those who are acted upon. Without this grammatical device, primates cannot describe a causal relation between events, or express a proposition like "Sylvia is trying to handle Martha's infant, but Martha is running away," or "Fat Tony challenged Halliburton and was wounded."

Because they do not create rule-governed call combinations, primate speakers cannot express much more than what can be expressed by a single vocalization. An infant grunt expresses a willingness to interact in a friendly way, and a threat-grunt simultaneously expresses both aggression (to the opponent) and a request for support (to potential allies). The lack of syntax, moreover, lends ambiguity to call meaning. A male baboon's alarm wahoo, for instance, cannot really be described as a command to action (*"Run into the trees"*) because not all baboons run into trees upon hearing the call, and baboons already in trees will also give this call if they see a lion. Nor is the call simply a noun (*"Lion"* or *"Carnivore"*), because it consistently evokes a flight response from at least some listeners. Instead, the baboon's alarm wahoo seems best described as a proposition: a single utterance or thought that simultaneously incorporates a subject and a predicate.

Baboon alarm calls, like those of other primates, are thus holistic utterances, simultaneously both eventish and objectish because they incorporate both reference to an object and a disposition to behave toward that object in a particular way. But there is no evidence that an alarm wahoo to a lion can be modified to elaborate upon the characteristics of the lion in question. Through repetition and changes in amplitude, alarm calls may inform others of the immediacy of danger, but they cannot specify whether a lion is big or small, sleeping or stalking, or here today and gone tomorrow.

Call perception

Most of the features that make nonhuman primate communication so strikingly different from human language concern vocal production. Viewed from the speaker's perspective, primate vocalizations and human speech could hardly be more different. The picture changes, however, when we turn to animals' perception of calls and their assessment of call meaning.

Repertoire size, development, and learning

As we have already noted, baboons and other monkeys produce a small repertoire of acoustically fixed, species-specific calls that are closely tied to particular social contexts and change little during development or adulthood. By contrast, when it comes to perception and comprehension monkeys have a much larger repertoire and display an almost open-ended ability to learn new sound-meaning pairs throughout their lives. They also require experience before they can use calls in the appropriate context or respond appropriately to the calls of others.

Young vervet monkeys, for example, need several years before they can give each of their alarm calls to the appropriate predator, and during this period they exhibit the same sort of overgeneralization shown by human infants as they learn the meaning of words (Cheney and Seyfarth 1990). When vervet infants first start to give alarm calls, at around six months of age, they give alarms to many species, like pigeons, geese, and warthogs that pose no danger to them. But their mistakes are not entirely random. Infant vervets give leopard alarms primarily to large, terrestrial mammals, eagle alarms primarily to birds, and snake alarms primarily to long, snakelike objects—like the cables that run from our playback speakers. With time and experience they narrow the range of species that elicit alarm calls, giving leopard alarms only to mammalian carnivores, eagle alarms only to martial and crowned eagles, and snake alarms primarily to pythons. Vervet monkeys also require several months' experience before they begin to respond appropriately to other group members' alarm calls (Cheney and Seyfarth 1990).

In the cross-fostering experiments described in Chapter 10, we were able to make a direct comparison between production and perception. Whereas the cross-fostered juveniles continued to use their own species' calls in contexts where their peers produced another, their adoptive mothers and group-mates learned to recognize their calls. They behaved as if they were thinking, "There goes Jacquie again, *gruffing* when she should be *cooing*. Oh well." Perception and comprehension were clearly more flexible than production (Seyfarth and Cheney 1997).

Finally, in their natural habitats vervets, baboons, and other primates learn to recognize the alarm calls of other species like birds and ungulates, even though these calls are acoustically very different from their own (Cheney and Seyfarth 1985a; Hauser 1988). And throughout their lives they must continually learn to identify the voices of individuals who join their group or are born into it. Scientists who work with monkeys in laboratories often remark upon monkeys' ability to learn the voices of new caretakers, or to associate a sound like the jangling of keys or the beep of an electronic card-swiper with the imminent delivery of food. Primates seem, in other words, to have an almost open-ended ability to learn new sound-meaning pairs.

One of our acquaintances, Wayne Hansen, a farmer and wildlife conservationist in Namibia, had a pet baboon, Elvis, who acted as his assistant car mechanic. While Wayne lay on his back under his Land Rover, Elvis would sit near the toolbox, handing Wayne spanners, wrenches, and fuel filters as Wayne requested them. Wayne reports that Elvis had difficulty distinguishing "Number 10 spanner" from "Number 12 spanner." This problem was easily solved by Wayne saying "No, the other spanner," whereupon Elvis would try again. But of course Elvis never said "spanner" or "wrench" himself.

Like Elvis, subjects in the ape language projects have demonstrated an impressive ability to learn the association between a sign or a word and the object or event that it designates. Their production of signs, however, has often lagged behind. The same is true of syntax. In many of the ape language projects, subjects could be taught to comprehend even quite complex sentences, such as "Put the pink bear into the bucket and bring it here." But in no case could the same animals produce such a sentence (e.g., Terrace 1979; Savage-Rumbaugh 1986).

When it comes to repertoire size, development, and learning, therefore, the difference between production and perception is striking. Where call production is concerned, nonhuman primate vocalizations are strikingly different from language. In the realm of perception and cognition, however, continuities are more apparent.

These generalizations apply with equal force to other mammals, including especially dogs. Consider Rico, a border collie in Germany (Fig. 49). Rico's proud owner asserted that Rico had learned the words for more than 200 different toys, stuffed animals, and balls. To determine whether this was really the case, Juliane Kaminski, Josep Call, and Julia Fischer randomly grouped Rico's toys into 20 sets of 10 items each. While Rico waited with his owner in a separate room, the experimenters arranged one set of items in another room elsewhere in the house. Then they rejoined Rico and instructed the owner to ask Rico to bring two randomly chosen items (one after the other) from the experimental room. Rico retrieved a total of 37 out of 40 items correctly, suggesting that his vocabulary size was comparable to that of language-trained apes, dolphins, sea lions, and parrots (Kaminski et al. 2004).

Next the experimenters tested Rico's ability to engage in "fast mapping," the ability to learn a new word by a process of exclusion (Wilkin-

Figure 49. The border collie Rico retrieves a toy. Photograph by Renate Ritzenhoff.

son et al. 1998). They placed a novel item together with seven familiar items in the experimental room. They first asked Rico to bring them one or two familiar items. They then said a word that Rico had never heard before (say, *bling*) and asked Rico to retrieve it ("Fetch the *bling*"). Rico retrieved the novel item even in the first session, and he was correct in seven of ten sessions thereafter. Apparently, he was able to infer that the novel word was linked to the novel item based on a process of elimination, or exclusion learning, a process previously thought to be unique to human children. Four weeks after this initial and sole exposure to these new items, Rico remembered the word-object associations for the novel items (Kaminski et al. 2004).

Rico also used gaze and attention to guide word learning. In a subsequent experiment, Rico's owner showed him two novel toys and said a new word while staring at one of the toys. When Rico was then asked to retrieve the named toy, he chose the toy that his owner had looked at, not the one his owner had ignored (J. Fischer, personal communication). In this respect, Rico's behavior was somewhat like that of young children at the earliest stage of word learning, who use inferences about the focus of a speaker's attention to map words onto objects (Chapter 8).

Of course, Rico is a highly motivated dog, and some of his talent may be accounted for by his domestication. Border collies are herders that have undergone generations of artificial selection favoring individuals skilled at attending to the attention and communication of their human trainers. Nonetheless, Rico's skills in learning new words and amassing a large vocabulary illustrate several important points.

First, many animals besides primates have an almost open-ended ability to form associations between a sound (or sign) and an object. They seem to understand the principle that sounds can designate objects. Second, most of these species can probably learn by fast mapping and exclusion and, after just a few exposures, retain the information they have acquired for several weeks if not years (for similar data on sea lions, see Schusterman et al. 2002). Of course, word learning in dogs, baboons, sea lions, and other species is not identical to word learning in human children, who learn many thousands of words without explicit training or reinforcement, and whose mind-reading skills rapidly outstrip other animals' (Markman and Abelev 2004; Fischer et al. 2004b). Finally, data from Rico and many other species highlight again the differences between production and perception. Dogs, baboons, sea lions, and chimpanzees have an almost open-ended ability to learn new sound-meaning pairs, but they are highly constrained in their ability to produce new vocalizations. Rico's skills in word retrieval, Ahla's talent in recognizing the bleats of individual goats, and wild baboons' ability to recognize the voices of 80 or more different individuals are all impressive, but how much more impressed would we be if Rico had 200 different barks, Ahla could speak the names of 50 different goats, or baboons had different grunts for each member of their group?

Call meaning

The same generalizations hold when we consider the cognitive operations that underlie baboons' assessment of the meaning of single vocalizations. Again, consider alarm calls. When it comes to production, these are classic primate vocalizations: acoustically invariant, relatively unaffected by learning, and tightly linked to specific contexts. When it comes to perception and cognition, the picture is much more complicated.

Recall from Chapter 10 that Diana monkeys associate three acoustically different sounds with a leopard: a leopard's growl, a male Diana monkey's leopard alarm call, and a female Diana monkey's leopard alarm call (remember that males' and females' alarm calls sound very different). In a similar manner, they associate three acoustically different sounds with a crowned eagle: an eagle's shriek, a male Diana monkey's eagle alarm call, and a female Diana monkey's eagle alarm call. The call-meaning relationship in the listener's mind is interesting in several respects.

First, it constitutes an arbitrary association between a sound and the thing for which it stands. There is nothing about the sound of a monkey's leopard alarm call that sounds like a leopard, and nothing about the sound of a monkey's eagle alarm that would obviously link it to an eagle. In much the same way, there is nothing in the acoustic details of baboons' infant and move grunts that would help a listener learn that one is given to infants (or in reconciliation) while another is given to announce a group move.

Second, in the Diana monkey's mind each call's meaning is defined not just by its relation to an object in the world but also by its relation to other calls in the monkey's repertoire. A male's leopard alarm is similar in meaning to a leopard's growl and a female's leopard alarm, but different in meaning from all of three eagle-associated calls. In the monkey's mind there exists a kind of "semantic space" in which the three leopard-associated sounds are closely linked in one cluster, whereas the three eagle-associated sounds are closely linked in another.

This leads to a third conclusion, that primate calls are acoustic units linked to particular concepts. When a Diana monkey hears a male's leopard alarm she appears to form a mental representation of the call's meaning. Then, when she hears a leopard's growl coming from the same location shortly thereafter, she forms a second representation and compares the two calls on the basis of their meaning. Her response to the growl is based on this assessment. The female, in other words, has a concept—a kind of mental image—of a leopard. The concept can be activated by any one of three quite different sounds that are linked together based on their shared meaning (see, e.g., Cohen et al. 2006). The concept is also amodal or multimodal, involving a combination of visual and auditory information (Gil da Costa et al. 2004; Ghazanfar et al. 2005).

As another example, consider the phenomenon of individual recognition by voice, which has been amply demonstrated in many species and underlies many of our playback experiments. Baboons clearly recognize an individual like Sylvia from her voice alone, regardless of whether she is giving a grunt, a contact bark, or a threat-grunt, and regardless of whether she is grunting loudly or softly, or vocalizing in a calm or agitated manner (Fig. 50). Despite wide variation in the acoustic

Figure 50. Sylvia in her 23rd year. Photograph by Anne Engh.

cues that mark a call as Sylvia's, and the fact that the calls of one indi-vidual may grade acoustically into the calls of another, listeners still link each call with a unique individual in a discrete, categorical fash-ion. Individual recognition occurs in so many contexts, with so many vocalizations, that it is hard to escape the impression that listeners have a mental representation, or concept, of Sylvia as an individual. If mon-keys were human, we would call this a concept of *person*.

In sum, whereas call production in primates is relatively fixed, the cognitive mechanisms that underlie call perception are considerably more complex. Underlying primates' assessment of call meaning is a rich conceptual structure in which calls are linked both to objects and relations in the world and to other calls in the species' repertoire. When responding to calls, monkeys act as if they recognize individuals and have concepts like *leopard, eagle, close associate,* and so on. The contrast between impoverished production and rich, conceptually based percep-tion argues strongly against the view that a concept cannot be acquired unless it is instantiated in one's language (reviewed by Gleitman and Papafragou 2005). Monkeys and apes have many concepts for which they have no words.

Communication and the theory of mind

The discontinuities between production and perception result in an oddly unbalanced form of communication: monkeys (and other animals) can learn many sound-meaning pairs but cannot produce new words, and they understand conceptual relations but cannot attach labels to them (Cheney and Seyfarth 1998). At the same time, it is crucial to remember that listeners are also signalers. It seems unlikely that qualitatively different neural and cognitive mechanisms would underlie a baboon's behavior when he hears another's alarm call and, seconds later, when he gives an alarm call of his own—or, more often, when he hears an alarm call, sees the lion, and remains silent.

But if this assumption is true, it only makes the dichotomy between call production and call perception even more puzzling. Why should an animal that can learn to associate hundreds of sounds and symbols with objects and events find it so difficult to produce novel calls or create novel call combinations? To answer this question it is useful to contrast word learning in animals with word learning in very young children.

As we discussed in Chapter 8, children actively attend to the speaker's gaze and focus of attention when inferring the referent for a novel word (Baldwin 1993). If a person is looking at a cup when she says "Blicket," the child will assume that the name for the cup is blicket. Although there are precursors to these abilities in the social interactions and communication of monkeys and apes, they remain rudimentary (Chapter 8). Baboons recognize when calls are being directed at them and they may have some understanding of other individuals' intentions, but unlike children they do not seem to recognize that gaze and intention can be informative. And while monkeys and other animals can learn the relation between hundreds of words (or signs) and objects, the mechanisms underlying word acquisition seem to be based on associative contingencies and exclusion learning rather than mental state reasoning.

Similarly, when it comes to production, monkeys' vocalizations seem designed to influence other individuals' behavior rather than (as in children) to affect their attention or knowledge. Although monkeys vary their calling rates depending upon the presence and composition of their audience, they do not act deliberately to inform ignorant individuals, nor do they attempt to correct or rectify false beliefs in others or instruct others in the correct usage or response to calls.

A rudimentary theory of mind, therefore, is fundamental to word learning. The extremely rapid pace of word learning in very young children depends on their ability to recognize that when a parent uses a new word in the presence of a novel object, she *intends* to use the word as a label for the object. Without this ability, the rapid word learning that is essential for a large vocabulary simply cannot evolve, and vocal production is sharply constrained.

The failure of monkeys and perhaps also apes to recognize the mental mechanisms that underlie communication may also partially explain the absence of syntax in their vocalizations. As we have discussed, at least some of the sounds produced by monkeys are functionally referential. Because they lack syntactic properties, however, their exact meaning is imprecise. Furthermore, baboons and other monkeys cannot elaborate upon the information contained in an alarm call. They cannot specify, for example, where a lion is, what it is doing, or when it was seen.

The lack of syntax in nonhuman primate vocalizations cannot be traced to an inability to recognize argument structure—to understand that an event can be described as a sequence in which an agent performs some action on an object. Baboons clearly understand the difference between *Sylvia threatens Hannah* and *Hannah threatens Sylvia*. Nor does the lack of syntax arise because of an inability to mentally represent descriptive modifiers (a *big* leopard as opposed to a *small* one) and prepositions that specify locations (a leopard *in* a tree as opposed to one *under* the tree). Captive dolphins (Herman et al. 1993a), sea lions (Schusterman and Krieger 1986), and African gray parrots (Pepperberg 1992) can be taught to understand and in some cases even to produce modifiers. In their natural behavior, therefore, nonhuman primates and other animals certainly act as if they are capable of *thinking,* as it were, in sentences. But the ability to think in sentences does not lead them to *speak* in sentences—in our view, because their lack of a theory of mind causes them not to understand what others might need to know. Because monkeys cannot distinguish between what they know and what others know, they fail to recognize that ignorant individuals must have events explained to them. As a result, even though they may mentally tag events as argument structures (who does what to whom), they fail to map these tags onto a communicative system in any stable or predictable way.

To summarize, the communication of nonhuman animals lacks three features that are abundantly present in the utterances of young children: a rudimentary ability to attribute mental states different from their own to others, the ability to generate new words, and syntax. We suggest that the absence of all three features is not accidental, and that the lack of one (a theory of mind) may explain the lack of the others (new words and syntax). Because they cannot attribute mental states like ignorance to one another and are unaware of the causal relation between behavior and beliefs, monkeys and perhaps also apes do not actively seek to explain or elaborate upon their thoughts. As a result, they are largely incapable of inventing new words and of recognizing when thoughts should be articulated. They do not understand the need to specify whether a leopard is *in* a tree or *on* the ground, nor do they comment about things in their absence (*"Do you remember the flea bag that ate my mother? Well, I saw it again yesterday"*). Instead, monkeys' calls reflect the knowledge the signaler has rather than the knowledge the signaler intends his audience to acquire.

A thoughtful reader may object to at least several parts of this proposal. What do we mean by stating that a rudimentary theory of mind is essential to word learning and production? After all, birds learn songs, but few would argue that song learning in birds is linked to a theory of mind. But the vocabularies of humans are more open-ended, more context-independent, and more functionally eclectic than the songs of birds. New sounds can be assigned to almost any object, event, or descriptor. Equally important, the adoption of new words within the community is rapid and reciprocal. Even very young children appear to have some tacit recognition of the relationship between comprehension and production. As a result, they understand that they can use a newly acquired word to influence another's behavior or attention. Similar capacities have yet to be demonstrated in the natural communication of any nonhuman animal.

Call combinations and the recognition of causality

Although the call sequences that monkeys produce are not structured and rule-governed, monkeys' perception of them is. Baboons hear different calls in combination all the time, and their assessment of them appears to rely heavily on rule-based expectations. Consider baboons' responses to call sequences mimicking a rank reversal (Chapter 6). Recall that in these experiments subjects heard the grunts of a low-ranking female (say, eight-ranking Shashe) combined with the fear barks of a higher-ranking female (say, third-ranking Beth). In the control sequence the anomalous sequence was maintained, but the grunts of a third female (say, alpha female Stroppy) who ranked higher than Beth were added. Subjects responded significantly more strongly to the sequences that appeared to violate the dominance hierarchy than to the control sequence.

Results like these suggest that, when baboons hear a sequence of calls, they associate each call with a particular individual, and each individual with a particular behavior and motivation. They also associate each call type with specific rules of delivery and evaluate its use in the current case in light of the callers' relationships. Listeners recognize, for example, that it is Shashe who is grunting, and that the grunt means that Shashe wants to interact in a friendly way with someone. They also recognize the fear barks as Beth's, and assume that Beth is nervous because she is interacting with a more dominant female. Most importantly, listeners recognize a causal relation between the calls. The pairing of Shashe's grunts with Beth's fear barks constitutes a violation of expectations only if listeners assume that Shashe is the one who is causing Beth to give fear barks.

Once we accept the notion of causality, it becomes clear that these simple sequences are similar in at least one respect to the words in a sentence: they are compositional. The individual calls preserve their meaning but the sequence as a whole conveys a meaning that is greater than the sum of its parts. Shashe's grunts retain their own meaning, as do Beth's fear barks. But taken together the two calls becomes a narrative about who is doing what to whom.

It also appears that in the control trial (Shashe and Stroppy both grunt and Beth fear barks) listeners are engaged in a simple form of parsing, sorting the three calls into their constituent groups in order to determine the meaning of the entire sequence. Steven Pinker (1994) illustrates the importance of parsing with the ambiguous sentence *On tonight's program Dr. Ruth will discuss sex with Dick Cavett*. The sentence is ambiguous because its constituent bracketing could be either [*discuss*] [*sex*] [*with Dick Cavett*] or [*discuss*] [*sex with Dick Cavett*].

In a simpler manner, the meaning of the control sequence is also ambiguous. Are Beth's fear barks caused by Shashe's or Stroppy's grunts? The sequence itself provides no clues about which causal relation is more important (Fig. 51). But listeners consistently responded as if their constituent bracketing included the causal relation [Stroppy's grunt → Beth's fear bark] rather than the causal relation [Shashe's grunt → Beth's fear bark], because subjects acted as if the sequence did not violate their

Figure 51. Lizzie listens to another female's grunt as her sister Lissa grooms her. How do baboons determine the causal factors that underlie each others' vocalizations? Photograph by Keena Seyfarth.

expectations. Upon hearing a series of calls whose meaning was ambiguous, therefore, subjects acted as if they had resolved the ambiguity in their minds by dividing the sequence into its constituent parts.

Finally, we take it for granted that humans interpret sentences in light of information we already possess about people and the world. The external knowledge we bring to the table when interpreting a sentence comes from our past experiences and our knowledge of how the world works. To appreciate the ambiguity of the sentence quoted above, it helps to know that Dr. Ruth is an advisor on topics related to sex, that Dick Cavett is (or was) a TV talk show host, and that while it is common to talk in public about sex in general, it is generally considered inappropriate to have public discussions about sex with a particular person.

In a similar manner, baboons display their knowledge and expectation about the world when they listen to call sequences. In the rank reversal experiments, listeners recognized individuals, assessed their motives and behavior, and evaluated the entire call sequence in light of what they knew about the dominance hierarchy in their group. Of course, none of these call sequences was produced by a single individual. In their manner of production, they could hardly be more different from language. Nonetheless, baboon listeners interpreted the meaning of a sequence of vocalizations in ways that are quite reminiscent of the cognitive operations that underlie human sentence processing.

The syntax of social knowledge

The vocalizations of monkeys clearly lack any properties that we would be tempted to call syntactic. Nevertheless, their social knowledge, their assessment of call meaning, and their parsing of call sequences display a number of syntactic properties.

First, knowledge is *representational*. When a monkey hears a vocalization she acquires information that is highly specific—about a particular sort of predator, or about a particular individual, her motivation to interact in specific ways with another, or the other animal's reaction.

Second, social knowledge is based on properties that have *discrete values* (Worden 1988). Sylvia is recognized as a high-ranking female member of the Selo matriline. Balo is also female, but she is low-ranking and a member of a different matriline. A move grunt is recognized as one call, associated with a particular motivation; an infant grunt is, in the information it conveys, a different signal.

Third, animals combine these discrete-valued traits to create a representation of social relations that is *hierarchically structured*. Baboons, for example, create a nested hierarchy in which others are placed in a linear rank order and simultaneously grouped according to matrilineal kinship in a manner that preserves ranks both within and across families (Chapter 6).

Fourth, social knowledge is *rule-governed and open-ended*. Baboons recognize that vocalizations follow certain rules of directionality that must, for instance, correspond to the current dominance hierarchy. Threat-grunts are given only by dominant animals to subordinates, fear barks are given only by subordinates to dominants, but infant and move grunts can be given in either direction. Knowledge is open-ended because new individuals can be added or eliminated without altering the underlying structure, and because the set of all possible interactions is very large (Worden 1998; Seyfarth and Cheney 2003). Taken together, the rule-governed and open-ended properties of primate social knowledge lead to a cognitive system that allows animals to comprehend a huge number of messages from a finite number of signals. If a baboon understands that *Sylvia threat-grunts and Hannah screams* carries a different meaning from *Hannah threat-grunts and Sylvia screams,* she can make the same judgment for all possible pairs of individuals in the group, including any new individuals who may join.

Fifth, knowledge is *propositional*. Baboons evaluate the meaning of call sequences in terms of other individuals' identities and motives and the causal relations that link one individual's behavior with another. That is, they represent in their minds (albeit in a limited way) the individuated concepts of "Sylvia," "Hannah," "threat-grunt," and "scream," and they combine these concepts to create a mental representation of one individual's intentions toward another. In so doing, they interpret a stream of sounds as a dramatic narrative: "Sylvia is threatening Hannah and causing her to scream."

Sixth, knowledge is *independent of sensory modality*. While playback experiments allow us to explore the structure of primates' social knowledge and demonstrate that such knowledge can be acquired through vocalizations alone, social knowledge is also obtained visually. Indeed, we now know that at the neurophysiological level visual and auditory information are integrated to form a multimodal representation of call meaning (Ghazanfar and Logothetis 2003; Gil da Costa et al. 2004; Ghazanfar et al. 2005).

These properties of nonhuman primates' social knowledge, while by no means fully human, bear important resemblances to the meanings we express in language, which are built up by combining discrete-valued entities in a structured, hierarchical, rule-governed, and open-ended manner. This leads to the hypothesis that the internal representations of language meaning in the human brain initially built upon our prelinguistic ancestors' knowledge of social relations (Cheney and Seyfarth 1998; Worden 1998). Indeed, as Worden (1998:156) argues, "no other candidate meaning structure has such a good fit to language meanings."

We are not suggesting that all of the syntactic properties found in language are present in primate social knowledge. Such a claim would be entirely unjustified, given the many features of language (syntactic features in particular) that have no counterpart in the communication or cognition of any nonhuman primate and that almost certainly evolved long after the divergence of the hominid line from the common ancestors of humans and chimpanzees (for recent discussions, see Jackendoff 1999; Calvin and Bickerton 2000; Hauser et al. 2002; Burling 2005; Johansson 2005). Instead, focusing on the early, prelinguistic stages of language evolution, we suggest that the precursor of the hominoid mind evolved in an environment characterized by social challenges and that such competition created selective pressures favoring structured, hierarchical, rule-governed intelligence. Because this social intelligence shares several features with language, many of the rules and computations found in human language may have first appeared as an elaboration of the rules and computations underlying social interactions.

Natural selection, social knowledge, and the cognitive precursors to language

We can now restate the social intelligence hypothesis as it applies to the cognitive precursors of language. In doing so, we assume that the generalizations we have drawn concerning the evolution of communication and cognition in baboons also apply to other monkeys and apes, perhaps to many group-living birds and mammals, and to the prelinguistic ancestors of modern humans.

In groups of long-lived, highly social animals, communication and cognition are linked to fitness. To survive, avoid stress, reproduce, and raise offspring who are themselves successful, individuals need both a system of communication that allows them to influence other animals' behavior and a system of mental representations that allows them to recognize and understand other animals' relationships. Because these mental representations concern animate creatures and are designed to predict behavior, they include information (if rudimentary) about other individuals' mental states, and about the causal relations between one social event and another. The demands of social life have thus led to the evolution of animals who represent their world not just as a collection of different individuals but as a collection of actors, each one of whom is predisposed to behave in certain ways toward each of her possible partners according to a number of (usually) predictable rules. Evolution has produced individuals who have propositional attitudes encoded in a *language of thought.*

The language of thought in our prelinguistic ancestors was adaptive in its own right. Perhaps more important, selection favoring a language of thought created individuals who, despite their limited vocal repertoires, were preadapted to developing language itself—preadapted because they already possessed many of the cognitive skills required for the understanding of words and sentences.

The hypothesis that social cognition acted as an initial prime mover in the evolution of language is not just another "just so story." To the contrary, it proposes an evolutionary scenario in which several of the cognitive precursors to language are directly linked to fitness. To survive and reproduce, primates (and perhaps other animals) must have a representation of sound-meaning relations that is based on a rich conceptual structure, and a representation of social relations that is discretely coded, combinatorial, hierarchically structured, rule-governed, and propositional. These skills are directly linked to reproductive success in baboons (see Chapter 5), and it seems reasonable to assume that they were also correlated with fitness in our prelinguistic ancestors.

The social origins hypothesis adds a slightly new wrinkle to theories of language evolution because it proposes that the precursors to language in the cognitive abilities of our prelinguistic ancestors had a grammatical flavor to them. In a widely cited hypothesis, Derek Bickerton has proposed that language evolved in two stages: first "proto-language" and then modern language (1990; Calvin and Bickerton 2000). Bickerton's model of proto-language is drawn from four sources: pidgin languages (Bickerton 1981), the language of individuals who have been isolated from adults during childhood (Curtiss 1977), children's language at the one-word stage; and the signing of captive apes (e.g., Savage-Rumbaugh 1986). In essence, proto-language is language without syntax (Jackendoff 1999, 2002). By contrast, the social origins hypothesis suggests that some of the cognitive operations that underlie modern syntax were among the earliest precursors of language. Specifically, before language appeared, natural selection favored individuals who, upon hearing a sequence of calls, could combine several discrete, meaningful elements in a rule-governed manner to create a complex, propositional representation of events. In Bickerton's hypothesis, proto-language is grammatically impoverished, making it difficult to imagine a gradual transition from proto-language to language. The social origins hypothesis may in some respects help to alleviate this problem.

The social origins hypothesis also makes chronological sense. If we assume that social complexity favored increasingly sophisticated cognitive abilities, we can imagine how these skills might have created an environment in which natural selection favored more flexible articulation, a full-blown theory of mind, the ability to generate new words, and the ability to create sentences. By contrast, it is difficult to imagine how—or why—these uniquely human skills would have evolved if humans had not first possessed the conceptual capacity that made them adaptive.

Indeed, if one accepts the striking parallels between social cognition and the mechanisms that encode meaning in language, and agrees that the former is a generalized primate trait while the latter are unique to humans, it is hard to imagine that the earliest forms of human syntax did *not* build upon these preexisting cognitive skills. Before hominids produced syntactic utterances, they assigned meaning to other individuals' calls and extracted syntactic, rule-governed, propositional information from the vocal interactions of others. Language-like perception and cognition thus preceded and set the stage for language-like production.

First thought, then language

We conclude, then, that long before our ancestors spoke in sentences they had a language of thought in which they represented the world—and the meaning of call sequences—in terms of actors, actions, and those who are acted upon. Long before they could engage in the computations that underlie modern grammar they performed the computations needed to understand their societies. As a result, the discrete, compositional structure we find in spoken language did not first appear there. It arose, instead, because understanding social life and predicting others' behavior requires discrete, compositional thinking. And the propositions that are expressed in language did not originate with language. They arose, instead, because to succeed in a social group of monkeys or apes one must understand an elementary form of propositional relations. The linguistic revolution occurred when we began to express this tacit knowledge, and to use our cognitive skills in speaking as well as listening. The earliest syntactic utterances, however, were not entirely original. They described relations that their speakers already understood and had a formal structure that grew out of their speakers' knowledge of social relationships.

TWELVE

Baboon Metaphysics

We can thus trace causation of thought ... it obeys the same laws as other parts of structure. **CHARLES DARWIN, 1838:** *NOTEBOOK M*

In nature, there is no representation without evolution, and perhaps there is no evolution beyond a certain point without the capacity to represent the world. **PHILLIP JOHNSON-LAIRD, 1983:** *MENTAL MODES*

Darwin's goal was to link metaphysics with survival and reproduction. Baboons allow us to do this. Like the beak of a finch, the mind of a baboon has been shaped by natural selection—in the baboon's case, by natural selection acting in a social environment.

Baboons live in a society where reproductive success depends on social skills. To survive and reproduce, a male must live a long life, maintain high dominance rank, establish close (albeit temporary) bonds with females, and protect the infants he has fathered. Among males, longevity and lifetime rank appear to be the best predictors of reproductive success (Alberts et al. 2003), and fundamental to a male's lifetime rank is his ability to deal with other males. During their lives, males experience the greatest stress from predation and challenges to their status.

To achieve the same goals, a female must live a long life, raise healthy infants, protect them from infanticide, and maintain an extensive network of related and unrelated companions. Among females, longevity and infant survival are the best predictors of reproductive success, and the best predictor of infant survival is the extent of a female's social integration (Silk et al. 2003). During their lives, females experience the greatest stress from predation, especially when it involves the loss of a close companion, challenges to their infants from infanticidal males, and challenges to their family's status. When faced with the loss of a close companion, a female can alleviate stress by broadening and extending her social network; when her infant is threatened with infanticide she can alleviate stress through friendship with an adult male; and when confronted with a challenge to her family's dominance rank she can maintain her status through close bonds with kin (Beehner et al. 2005; Engh et al. 2006a,b).

Natural selection has favored in baboons (and, by extension, other monkeys) a mind that is specialized for observing social life, computing social relations, and predicting other animals' behavior. The brains of monkeys contain areas that are particularly sensitive to other individuals' orientations, movements, gaze direction, and intentions (Chapters 6 and 8). Their communication is equally specialized. When a female baboon hears another female's vocalization, she does not just hear a sound. She perceives a signal that evokes a representation of the caller, what she is doing, her rank and family membership. Baboons seem compelled to respond this way. Just as we cannot hear a word without thinking about its meaning, so baboons cannot hear a vocalization without thinking about the animal who is calling and the events the call describes. And they cannot hear an exchange of vocalizations between Sylvia and Hannah without thinking about these animals' identities, ranks, and family membership, about their relationship, and about its place in the social order. When baboons hear Sylvia, Hannah, or any other animals interacting, they respond instantly and, as far as we can tell, unconsciously. They have a social mind that is innately computational and judgmental.

But while the tendency to make social judgments and form a representation of call meaning may be innate, the *content* of these representations changes all the time. Baboons are always monitoring each other and keeping track of who is consorting with whom, who has fallen in rank, who is moving up, and which families are feuding with each other. Within hours of any societal change, they incorporate this new information into their expectations. They have an innately representational mind that is always open to new information.

Although we know much less about baboons' (and other primates') knowledge outside the social domain, there are at least two reasons to suspect that social life has had the most profound influence in making their minds different from the minds of other animals. First, social pressures are pervasive. It is obviously crucial that baboons have extensive knowledge of their home range, the spatial and temporal distribution of trees and other plants, and the behavior of predators. These challenges, however, are inextricably bound up in social life. To find her way about, select the right food, and avoid predation, a young baboon must observe and learn from others. Once she has done so, she must somehow deal with the fact that they, too, want to eat the same food and avoid the same predators. For gregarious animals, ecological challenges have social dimensions.

Second, social pressures are more dynamic than those faced outside the social domain. Once a baboon has mastered her environment, the job is pretty much done—or at least done as well as it can be. Mastering the problems posed by another baboon is quite a different thing, because once you think you understand Sylvia and can predict what she will do next, she changes her behavior—perhaps, perversely, in response to your own behavior. Ecological changes occur slowly, over weeks, months, and years. But social changes occur constantly, rapidly, and unpredictably. And an animal's response to these changes will provoke further changes in others, which will elicit responses, and so on throughout the individual's life in a never-ending, reinforcing spiral.

The social world is inherently dynamic. By contrast, the physical world—at least for baboons—is comparatively static. Unlike humans, baboons do not look at their environment and ask "How can I change it?" They do not wonder about how they could make food more accessible, easier to process, or more frequently available. If they catch a baby impala they will eat it, but they never think about how they might devise a trap or a new hunting technique that would make catching an impala easier or more predictable. They lack the insight to imagine a different world.

Although their minds have been overwhelmingly shaped by social life, baboons have only a limited ability to recognize the mental states of others. Though they have at least a rudimentary recognition of other individuals' intentions and motives, they seem oblivious to others' knowledge and beliefs. As a result, whereas they make relatively complex inferences about the target of other individuals' calls and other individuals' motivations toward themselves, they do not go out of their way to inform others about what they know, even when the others lack crucial information. Baboons extract rich causal narratives from other animals' calls, but these narratives remain private. Unlike humans and even very young children, they feel no urge to gossip or share information.

Monkeys' access to their own thoughts is similarly limited. Although they appear to monitor their knowledge in some limited ways, they seem incapable of the sort of "what if" introspection that permits deliberate planning and the weighing of alternative strategies. And in the absence of introspection and mental time traveling, it is hard to imagine how any species could ever develop extensive tool use or culture. Baboons' social intelligence is impressive, but they live largely in the present tense. (By contrast, if you google "what if," you will receive over 150 million hits—an indication, perhaps, of too much of a good thing).

Equally striking is the difference in baboons' communication between comprehension and production. Baboons—like dogs, chimpanzees, and many other species—understand much more than they can say. Their language of thought is impressive; their ability to articulate their thoughts much less so.

These omissions lead to at least one general conclusion about the evolution of mind, communication, and society in animals. Baboons teach us that it is possible to have a complex society based on cognitive processes that are both computational and representational without either language or a theory of mind. Concepts (of a sort) can exist without words; computation can occur without grammar. Along with many other species of animals, baboons provide us with a natural experiment that allows us to ask "What is thought—what can it possibly be—without language and a theory of mind"?

As Darwin hoped it would, the study of baboon metaphysics may help us understand the evolution of thought, communication, and language in humans and other species. We can now put his comparative method to the test and begin to trace the causation of thought.

Comparing baboons and nonprimate species

How do baboons differ from nonprimate species? At this stage in our work, the simple answer is: we're not sure. Many birds and mammals, particularly those living in large social groups, have a repertoire of calls that they use to manipulate other individuals' behavior and a language of thought that allows them to recognize and predict social events. We now know that many birds and nonprimate mammals recognize each other's dominance ranks. Some may also recognize other individuals' social relationships. In the laboratory, parrots, dogs, sea lions, and dolphins have also learned to associate arbitrary sounds with objects and actions, as if they have a rudimentary understanding of the function of nouns, modifiers, and verbs. They have the necessary mental architecture because their social and physical environment forces them to represent the world in terms of "nouns" (other group members, predators, food), "modifiers" (the ripeness of fruit, the size of prey), and "verbs" (what is happening). By many measures there are few cognitive differences between monkeys and other animals.

Baboons (and, by extension, other monkeys) may differ from other animal species in a few, though as yet untested, respects (Chapter 7). They may be superior at classifying individuals along multiple, hierarchical dimensions and at placing them in overlapping categories. Their causal inferences in the social domain may also be more indirect and complex. Similarly, whereas other animals predict their companions' behavior on the basis of learned contingencies, baboons may do so additionally by attributing motives and intent. But these differences, if they exist at all, may not represent a startling, quantum leap in cognition. Baboons have an elaborate society and sophisticated social intelligence, but so may many other birds and mammals.

The differences between baboons and other species may lie not so much in their innate tendency to acquire social knowledge as in the particular details of the knowledge they acquire. Baboon society is organized around a linear rank order of matrilineal families. As a result, the computations that underlie their social knowledge are based largely on kinship and rank, and the picture that emerges in a baboon's mind can be diagrammed as a branching, hierarchical tree. Within a baboon group, some relationships are enduring, others temporary. Some are marked by close spatial proximity, others are not. Some relationships can be predicted on the basis of transitivity; others cannot. Some rank relations change frequently, others remain static for years at a time. Many songbirds, on the other hand, live in monogamous territorial pairs and need assess only their neighbors' relative competitive ability. As a result, their social categories and the computations that underlie them may be somewhat simpler. A greater challenge will be to determine the degree of similarity in the languages of thought between very different species whose social lives appear, at least superficially, to be equally complex. Pinyon jays and hyenas live in social groups that are ostensibly similar to baboons'. Do these species also classify others simultaneously according to multiple criteria? Do they recognize that different types of social relationships are characterized by different rules? When they hear a vocalization, do they make the same causal inferences that baboons do? If so, are these inferences also based on considerations like the nature of recent events and the relationships that exist among other group members? The comparative experiments to answer these questions have not yet been conducted.

Comparing baboons, chimpanzees, and humans

Roughly 30 million years ago, baboons, chimpanzees, and humans shared the same ancestor. The ancestral line leading to baboons and other Old World monkeys then diverged. For almost 20 million years thereafter, chimpanzees and humans shared a common ancestor before separating roughly five to seven million years ago. In what ways have baboon and chimpanzee minds diverged since their separation? And what selective pressures might have resulted in the obvious differences between the chimpanzee and human minds?

The most striking differences between baboons and chimpanzees can be found in brain size and structure, social organization, and tool use. As we noted in Chapter 7, baboon brains are smaller than those of chimpanzees, both absolutely and relative to body size. Chimpanzee brains also differ qualitatively from those of Old World monkeys, with comparatively greater elaboration of the cerebellum and frontal lobes (Rilling 2006).

Baboons live in groups of up to 100 individuals who move together as a unit and are organized around a core of genetically related females. Chimpanzees, in contrast, live in large fission-fusion communities of up to 40 males, some of whom are related, and a varying number of usually unrelated females. Within a chimpanzee community, individuals join and separate in a largely unpredictable manner, although males are more likely to form temporary foraging and patrolling parties while females are more likely to travel alone (Goodall 1986; Watts and Mitani 2001; Boesch et al. 2002). Although little is known about chimpanzee communication in the wild, chimpanzees appear to have a vocal repertoire that is roughly the same size as baboons', and their vocal development seems equally constrained. Throughout Africa, chimpanzees use a wide variety of tools, including stones to break open nuts, small sticks and stems of grass to fish for ants and termites, and large sticks as clubs and hammers (Chapter 8). Baboons, in contrast, have never been reported to use tools in the wild. Indeed, the New World capuchin monkeys are the only monkeys that regularly use tools under natural conditions, though there is no evidence that they modify or prepare tools in advance of their use.

Chimpanzees' knowledge of other individuals' social relationships is hard to compare with monkeys', because no study has yet asked a chimpanzee what she knows about the ranks, kinship, or friendships of other chimpanzees. In the laboratory, chimpanzees outperform monkeys in tests that require tool use, planning, and imitation, but, at least in some contexts, monkeys are also capable of a limited degree of imitation. Chimpanzees select tools in advance of their use, suggesting that they can, to some degree, introspect about their future intentions and goals. Monkeys, though, are capable of monitoring their knowledge to a limited degree. Chimpanzees may be better than monkeys at reading other individuals' intentions, goals, and visual perspectives, but the evidence here is by no means clear. Like monkeys, chimpanzees show little evidence of teaching. They experience grief but their capacity for empathy remains controversial. It even remains unclear whether chimpanzees differ qualitatively from dogs and some birds in their ability to read their own and others' minds (Chapters 8 and 9). They may well do so, but the evidence to date is far from definitive.

Fission-fusion societies like those found in chimpanzees may place strong selective pressure on the ability to imagine and plan hypothetical social interactions with individuals who may not be encountered for several days or even weeks. They may also favor the ability to represent oneself and others in different places, times, and scenarios (Dennett 1987b; Barrett et al. 2003). If true, chimpanzees' social structure may have favored a language of thought that includes a form of "what if" episodic memory that is qualitatively different from that found in monkeys. It remains for future research to test this hypothesis.

The human mind

Whatever the outcome of this work, it remains indisputable that the human mind differs qualitatively from that of other apes. In the few million years after the divergence of the hominid line, brain size increased dramatically, including in particular the prefrontal and temporal association cortices—the areas associated with complex cognition (Deacon 1997; Rilling 2006). We suspect that, as our uniquely human traits began to appear, the continuing evolution of a theory of mind played the crucial role as catalyst and prime mover, facilitating and leading to the evolution of all of the traits that are uniquely human, including speech, teaching, elaborate tool use, and culture.

Consider word learning in children, a skill that depends crucially on the listener's ability to make inferences about a speaker's intent. If a person is looking at a particular object when she says an unfamiliar word, a child assumes implicitly that the word is a label for that object. Monkeys and apes appear to be much less adept at recognizing the informative content of attention.

Or take syntax, another uniquely human trait that is adaptive because it allows individuals to communicate an infinite number of messages using a finite number of sounds. Syntax overcomes the limitations imposed by a limited vocabulary, and it allows individuals to communicate information about relations between objects or events. But for syntactic communication to be favored by natural selection, the speaker must first understand that his audience *needs* to have events or ideas explained to them, and he must recognize that speech can be used to inform, warn, cajole, and deceive.

Once an individual can represent another person's mental state and compare his own mental states to others', he can begin to think (and speak) about mental processes and content: he can begin to think or say that another individual "believes that...," "thinks that...," or "wants that...." Philosophers call this *intentionality,* or "aboutness." Mental states are always *about* other things: baboons are baboons, but a belief is always a belief *about* something. You can hold a belief about baboons that is true or false or wildly delusional, but baboons themselves cannot be true, false, or wildly delusional, except in a metaphorical sense. Mental states are also recursive, because nested inside the mental state is its content. If you believe that Ian thinks that his boat has just run over a hippo, you assume that Ian has thoughts and that, at this instant, one of them contains a representation of a hippo. If you can examine your own mental states, you know that you have thoughts and that, at this moment, one of them contains a representation of Ian's thought that contains a representation of a hippo.

Representing another individual's mental state is thus inherently, automatically recursive, and representing your own thoughts about another person's mental state is doubly so. When evolution favors a theory of mind, therefore, it necessarily favors recursive thinking, because there is no way to represent the content of another's thoughts without nesting the content inside of the thought. And when evolution favors the ability to express your knowledge about another's mental state it automatically favors recursive speaking, because recursion provides an excellent way to express the relation "thinks that...," "wants that...," "believes that...."

Modern languages exhibit recursion regardless of whether the sentence does or does not concern mental state attribution. An example of the former is "She thinks that he doesn't know that she likes him"; an example of the latter is "This is the cat that ate the mouse that lived in the house that Jack built." We suggest that recursive *thinking* first appeared in the social and technological knowledge of hominids who could gain a reproductive advantage by representing the mental state of others (and themselves), and that recursive *speaking* appeared subsequently, pushed by the need to express the intentionality inherent in mental state attribution. From these modest beginnings, recursion became a pervasive component of human thinking and syntax.

It has been hypothesized that humans differ from other apes not only in the sophistication of their theory of mind but also in their motivation to share their intentions, emotions, and knowledge (Tomasello et al. 2005). Even very young children with only an implicit understanding of other people's minds are strongly motivated to share their ideas and empathize with others. Chimpanzees do cooperate with one another, and they may also make inferences about others' goals and motives when engaged, for example, in collaborative hunts. However, they do not appear to be as motivated as young children to cooperate and empathize with others. They seem much less sensitive than young children to the psychological mechanisms underlying cooperation. This motivation to share ideas and emotions with others almost certainly played a crucial role in the evolution of language, tool use, and culture. We will leave speculation about the selective pressures that might have favored high levels of cooperation in early hominids to others.

Because humans and chimpanzees differ strikingly in their theories of mind and in their motivation to communicate what they know, we suggest that the evolutionary pressures favoring individuals' ability to represent each other's knowledge—the rudiments of which we see in modern chimpanzees—created strong selective pressures favoring an ability to *express* this knowledge to others. In other words, having a theory of mind favored an ability to speak, expand one's vocabulary, and combine words in sentences to convey novel meanings. Thought came first; speech and language appeared later, as its expression.

This view, that a theory of mind and the motivation to share knowledge served as the driving forces behind the evolution of flexible vocal production, is consistent with recent genetic discoveries, which suggest that genetic changes leading to the evolution of flexible, modern phonation occurred at a relatively late date, after the human line had diverged from the common ancestor of human and chimpanzee. For example,

Wolfgang Enard, Svante Päabo, and their colleagues have identified a gene complex called FOXP2 that affects the flexibility of orofacial movement and hence vocal production. The gene differs between humans, on the one hand, and all other mammals including chimpanzees on the other. Genetic changes in the human line have thus occurred since the human-chimp split, and appear to have been the result of selection rather than any other mechanism. The changes were adaptive because of the benefits they brought to their possessors in the domain of speech and language (Enard et al. 2002). But as Päabo (2003) notes, changes in the genetic mechanisms underlying speech production would have been favored by natural selection *only* if they occurred in creatures that were already capable of a sophisticated form of communication and cognition. Just as they drove the evolution of other uniquely human traits, a theory of mind and the urge to share knowledge with others drove the evolution of flexible vocal production.

Finally, we return to technology—the use and manufacture of tools—a topic that has played little role in our discussion of baboon metaphysics. The reason, by now, should be obvious: baboons' adaptive specialization is their social intelligence; their technological skills are decidedly underwhelming. Baboons are also not motivated to change their physical world. Their brains are smaller than those of some great apes that live in smaller groups but use and manufacture tools. Indeed, there is a significant positive correlation between innovation, tool use, and brain size in primates, but not between innovation, tool use, and group size (Chapter 7). This has led to speculation that innovation and technology, not the demands of social life, have driven the evolution of large brains in primates (Reader and Laland 2002; Reader 2003). The hypothesis is persuasive, especially because we cannot yet point to a specific way in which monkeys' social knowledge differs qualitatively from that of other animals.

The ability to reflect actively upon one's own thoughts and beliefs permits the sort of introspection and mental time traveling essential not only for manipulating other individuals but also for manipulating things. The inventor of a tool must be able to imagine the tool's function in advance of its use and plan its manufacture accordingly. Furthermore, a tool's propagation requires that others recognize its use and understand that they may have to seek assistance from a knowledgeable tutor if they are to use it effectively. Like speech, tool manufacture and teaching have obvious adaptive values, but they require as a necessary precursor the ability to represent both another individual's and one's own thoughts and beliefs. It is difficult to imagine how any of these traits could have evolved completely independently of the others—difficult to imagine a hominid who could introspect about the invention of a new tool but was unable to recognize whether a pupil was ignorant or knowledgeable, or one who could inform an ignorant pupil but was incapable of flexible communication.

Baboons and other monkeys rarely if ever manufacture tools, but they do seem to have a limited capacity to monitor their own knowledge and to attribute simple mental states to others (Chapters 8 and 9). It therefore seems probable, as Jolly (1966) first proposed, that the technological and innovative skills evident in rudimentary form in chimpanzees (and hyperbolically so in humans) have their roots in the selective forces that originally favored the evolution of social skills. Although innovation, tool use, and technological invention may have played a crucial role in the evolution of ape and human brains, these skills were probably built upon mental computations that had their origins and foundations in social interactions.

Appendix

The matrilineal families of C troop, in descending rank order, in June 2006. Offspring who died before July 1992 are not shown. Females are represented by circles, males by triangles. Dead animals are crossed out. Males who emigrated from the group are depicted by an arrow. Two females who produced only sons are not included.

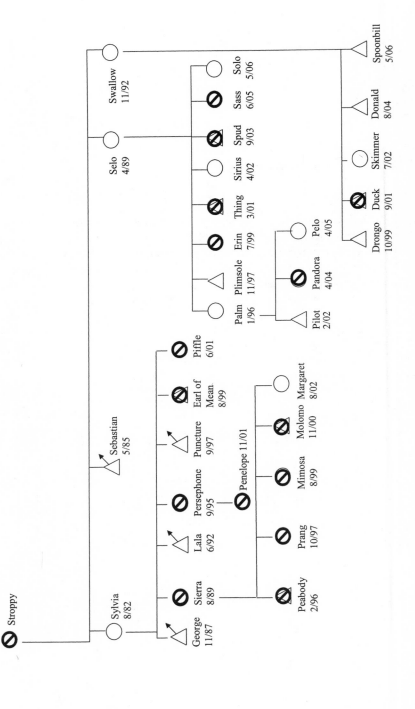

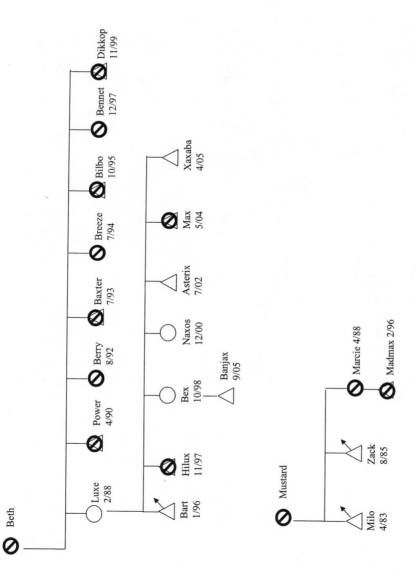

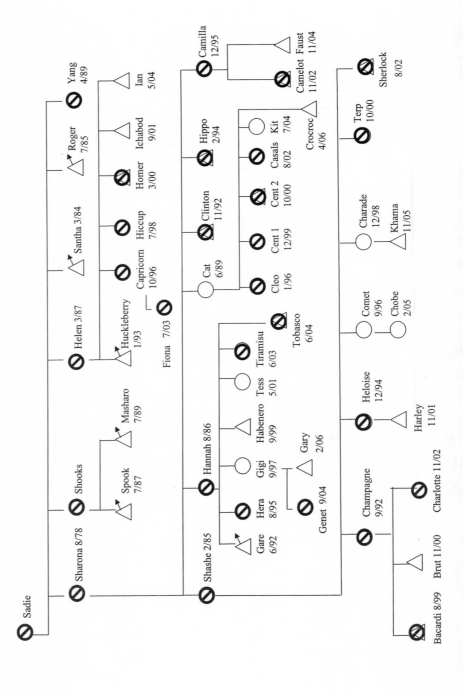

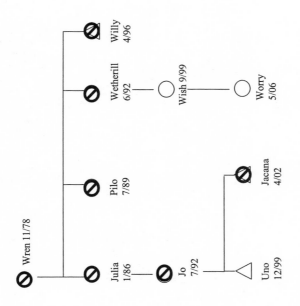

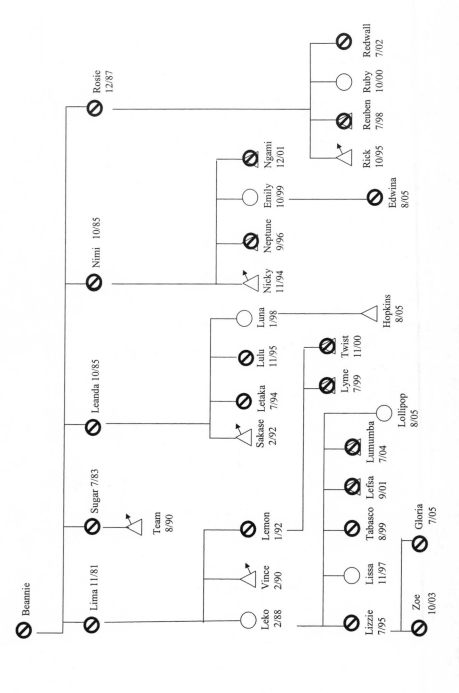

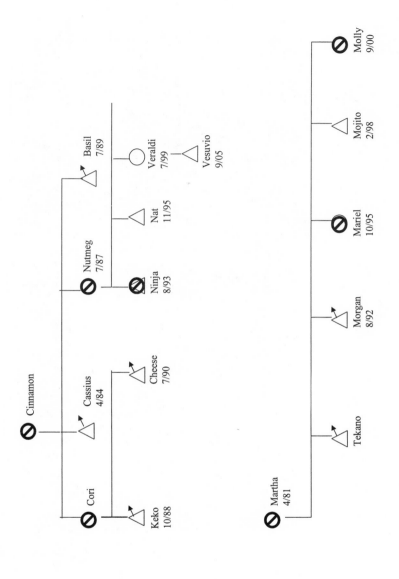

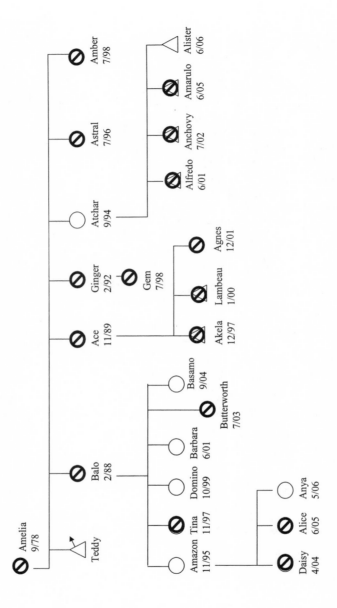

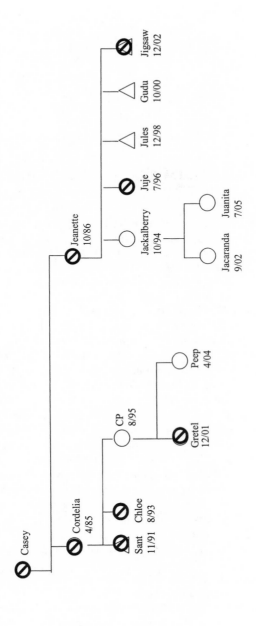

References

Abbott, D. H., Keverne, E. B., Bercovitch, F. B., Shively, C. A., Mendoza, S. P., Saltzman, W., Snowdon, C. T., Ziegler, T. E., Banjevic, M., Garland, T. J., and R. M. Sapolsky. 2003. Are subordinates always stressed?: A comparative analysis of rank differences in cortisol levels among primates. *Horm. Behav.* 43:67–82.

Adolphs, R., Russell, J. A., and D. Tranel. 1999. A role for the human amygdala in recognizing emotional arousal from unpleasant stimuli. *Psych. Sci.* 10:167–171.

Alberts, S. C., Sapolsky, R. M., and J. Altmann. 1992. Behavioral, endocrine, and immunological correlates of immigration by an aggressive male into a natural primate group. *Horm. Behav.* 26:167–178.

Alberts, S. C., Watts, H. E., and J. Altmann. 2003. Queuing and queue-jumping: Long term patterns of reproductive skew among male savannah baboons. *Anim. Behav.* 65:821–840.

Alexander, J. E. 1838. *An expedition of discovery into the interior of Africa, through the hitherto undescribed countries of the Great Namaquas, Boschmans, and Hill Damaras. Performed under the auspices of Her Majesty's government, and the Royal geographical society; and conducted by Sir James Edward Alexander.* H. Colburn, London.

Altmann, J. 1974. Observational study of behavior: Sampling methods. *Behaviour* 49:229–267.

Altmann, J. 1980. *Baboon Mothers and Infants.* Harvard University Press, Cambridge, MA.

Altmann, J., and S. C. Alberts. 2003. Variability in reproductive success viewed from a life-history perspective in baboons. *Am. J. Hum. Biol.* 15:401–409.

Altmann, J., Alberts, S. C., Haines, S. A., Dubach, J., Muruthi, P., Coote, T., Geffen, E., Cheeseman, D. J., Mututua, R. S.,

Saiyelel, S. N., Wayne, R. K., Lacy, R. C., & M. W. Bruford. 1996. Behavior predicts genetic structure in a wild primate group. *Proc. Nat. Acad. Sci. USA* 93:5797–5801.

Anderson, J. R., Montant, M., and D. Schmitt. 1996. Rhesus monkeys fail to use gaze direction as an experimenter-given cue in an object choice task. *Behav. Proc.* 37:47–55.

Andersson, C. J. 1856. *Lake Ngami; or Explorations and Discoveries during Four Years' Wandering in the Wilds of Southwestern Africa.* Harper and Brothers, New York.

Andersson, M. 1994. *Sexual Selection.* Princeton University Press, Princeton.

Apperly, I. A., Samson, D., and G. W. Humphreys. 2005. Domain-specificity and theory of mind: Evaluating neuropsychological evidence. *Trends Cog. Sci.* 9:572–577.

Arcadi, A. C., Robert, D., and F. Mugurusi. 2004. A comparison of buttress drumming by male chimpanzees from two populations. *Primates* 45:135–139.

Ardrey, R. 1969. Introduction to *The Soul of the Ape,* Eugene Marais. Penguin Books, Harmondsworth.

Arnold, K., and K. Zuberbuhler. 2006. Language evolution: Compositional semantics in primate calls. *Nature* 441:303.

Ashley Montagu, M. F. 1941. Knowledge of the ancients regarding the ape. *Bull. Hist. Med.* 10:530–543.

Asquith, P. J. 1995. Of monkeys and men. Pp. 309–326 in R. Corbey and B. Theunissen, eds., *Ape, Man, Apeman: Changing Views since 1600.* Leiden University Press, Leiden.

Astington, J. W., Harris, P. L., and D. R. Olson, eds. 1988. *Developing Theories of Mind.* Cambridge University Press, Cambridge.

Atance, C. M., and D. K. O'Neill. 2005. The emergence of episodic future thinking in humans. *Learn. Motiv.* 36:126–144.

Attwell, D., and S. B. Laughlin. 2001. An energy budget for signaling in the grey matter of the brain. *J. Cerebr. Blood Flow Metab.* 21:1133–1145.

Aureli, F., Cozzolino, R., Cordischi, C., and S. Scucchi. 1992. Kin-oriented redirection among Japanese macaques: An expression of a revenge system? *Anim. Behav.* 44:283–291.

Aureli, F., Das, M., and H. C. Veenema. 1997. Differential kinship effect on reconciliation in three species of macaques (*Macaca fascicularis, M. fuscata,* and *M. sylvanus*). *J. Comp. Psych.* 111:91–99.

Aureli, F., and F. B. M. de Waal, eds. 2000. *Natural Conflict Resolution.* University of California Press, Berkeley.

Aureli, F., Preston, S., and F. B. M. de Waal. 1999. Heart rate responses to social interactions in free-moving rhesus macaques (*Macaca mulatta*): A pilot study. *J. Comp. Psych.* 113:59–65.

Bachorowski, J. A., and M. J. Owren. 1995. Vocal expression of emotion: Acoustic properties of speech are associated with emotional intensity and context. *Psych. Sci.* 6:219–224.

Balda, R. P., and A. C. Kamil. 1992. Long-term spatial memory in Clark's nutcracker, *Nucifraga columbiana*. *Anim. Behav.* 44:761–769.

Baldwin, D. 1993. Infants' ability to consult the speaker for clues to word reference. *J. Child Lang.* 20:395–418.

Barraclough, N. E., Xiao, D., Baker, C. I., Oram, M. W., and D. Perrett. 2005. Integration of visual and auditory information by superior temporal sulcus neurons responsive to the sight of actions. *J. Cog. Neurosci.* 17:377–391.

Barrett, G. M., Shimizu, K., Bardi, M., Asaba, S., and A. Mori. 2002. Endocrine correlates of rank, reproduction, and female-directed aggression in male Japanese macaques (*Macaca fuscata*). *Horm. Behav.* 42:85–96.

Barrett, K. C. 2005. The origins of social emotions and self-regulation in toddlerhood: New evidence. *Cog. Emot.* 19:953–979.

Barrett, L., and P. Henzi. 2002. Constraints on relationship formation among female primates. *Behaviour* 139:263–289.

Barrett, L., and P. Henzi. 2005. The social nature of primate cognition. *Proc. Roy. Soc. Lond. B* 272:1865–1875.

Barrett, L., Henzi, P., and R. Dunbar. 2003. Primate cognition: From "what now?" to "what if?" *Trends Cog. Sci.* 7:494–497.

Barrett, P. H., Gautrey, P. J., Herbert, S., Kohn, D., and S. Smith, eds. 1987. *Charles Darwin's Notebooks, 1836–1844*. Cornell University Press, Ithaca, NY.

Barsalou, L. 2005. Continuity of the conceptual system across species. *Trends Cog. Sci.* 9:309–311.

Barsalou, L., Simmons, W. K., Barbey, A. K., and C. D. Wilson. 2003. Grounding conceptual knowledge in modality-specific systems. *Trends Cog. Sci.* 7:84–91.

Barth, J., Reaux, J. E., and D. J. Povinelli. 2005. Chimpanzees' (*Pan troglodytes*) use of gaze cues in object-choice tasks: Different methods yield different results. *Anim. Cog.* 8:84–92.

Barton, R. A. 1998. Visual specialization and brain evolution in primates. *Proc. Roy. Soc. Lond. B* 265:1933–1937.

Barton, R. A., and R. Dunbar. 1997. Evolution of the social brain. Pp. 240–263 in A. Whiten and R. W. Byrne, eds., *Machiavellian Intelligence II: Extensions and Evaluations*. Cambridge University Press, Cambridge.

Bartsch, K., and H. M. Wellman. 1995. *Children Talk about the Mind*. Oxford University Press, New York.

Bednekoff, P. A., and R. P. Balda. 1996. Observational spatial memory in Clark's nutcrackers and Mexican jays. *Anim. Behav.* 52:833–839.

Beehner, J. C., and P. L. Whitten. 2004. Modifications of a field method for faecal steroid analysis in baboons. *Physiol. Behav.* 82:269–277.

Beehner, J. C., Bergman, T. J., Cheney, D. L., Seyfarth, R. M., and P. L. Whitten. 2005. The effect of new alpha males on female stress in free-ranging baboons. *Anim. Behav.* 69:1211–1221.

Beehner, J. C., Bergman, T. J., Cheney, D. L., Seyfarth, R. M., and P. L. Whitten. 2006. Testosterone predicts future dominance rank and mating activity among male chacma baboons. *Behav. Ecol. Sociobiol.* 59:469–479.

Bergman, T. J., Beehner, J. C., Cheney, D. L., and R. M. Seyfarth. 2003. Hierarchical classification by rank and kinship in baboons. *Science* 302:1234–1236.

Bergman, T. J., Beehner, J. C., Cheney, D. L., Seyfarth, R. M., and P. L. Whitten. 2005. Correlates of stress in free-ranging male chacma baboons, *Papio hamadryas ursinus*. *Anim. Behav.* 70:703–713.

Bickerton, D. 1981. *Roots of Language*. Karoma, Ann Arbor, MI.

Bickerton, D. 1990. *Language and Species*. University of Chicago Press, Chicago.

Bloom, P. 2003. *How Children Learn the Meaning of Words*. MIT Press, Cambridge, MA.

Bloom, P. 2004. *Descartes' Baby: How the Science of Child Development Explains What Makes Us Human*. Basic Books, New York.

Bloom, P., and L. Markson. 1998. Capacities underlying word learning. *Trends Cog. Sci.* 2:67–73.

Blumstein, D. T., and K. B. Armitage. 1997. Alarm-calling in yellow-bellied marmots. I. The meaning of situationally variable alarm calls. *Anim. Behav.* 53:143–171.

Blyth, E. 1837. On the psychological distinctions between man and all other animals; and the consequent diversity of human influence over the inferior ranks of creation, from any mutual and reciprocal influence exercised among the latter. *Magazine of Natural History* 1:1–141.

Boas, G. 1948/1966. *Essays on Primitivism and Related Ideas in the Middle Ages*. Octagon Books, New York.

Boesch, C. 1991. Teaching in wild chimpanzees. *Anim. Behav.* 41:530–532.

Boesch, C. 2002. Cooperative hunting among Tai chimpanzees. *Hum. Nat* 13: 27–46.

Boesch, C., and H. Boesch. 1984. Mental maps in chimpanzees: An analysis of hammer transports for nut cracking. *Primates* 25:160–170.

Boesch, C., Hohmann, G., and L. Marchant. 2002. *Behavioral Diversity in Chimpanzees and Bonobos*. Cambridge University Press, Cambridge.

Boinski, S., and P. A. Garber, eds. 2000. *On the Move: How and Why Animals Travel in Groups*. University of Chicago Press, Chicago.

Bond, A. B., Kamil, A. C., and R. P. Balda. 2003. Social complexity and transitive inference in corvids. *Anim. Behav.* 65:479–487.

Bradbury, J., and S. L. Vehrencamp. 1998. *The Principles of Animal Communication*. Sinauer, Sunderland, MA.

Bräuer, J., Kaminski, J., Riedel, J., Call, J., and M. Tomasello. 2006. Making inference about the location of hidden foods: Social dog, causal ape. *J. Comp. Psych.* 120:38–47.

Brenowitz, E., and A. P. Arnold. 1986. Interspecific comparisons of the size of neural song control regions and song complexity in duetting birds: Evolutionary implications. *J. Neurosci.* 6:2875–2879.

Brenowitz, E., and D. Kroodsma. 1996. The neuroethology of bird song. Pp. 285–304 in D. E. Kroodsma and E. H. Miller, eds., *Ecology and Evolution of Acoustic Communication in Birds*. Cornell University Press, Ithaca, NY.

Bretherton, I. 1992. Social referencing, intentional communication, and the interfacing of minds in infancy. Pp. 57–78 in S. Feinman, ed., *Social Referencing and the Social construction of Reality in Infancy.* Plenum Press, New York.

Brodbeck, D. R. 1994. Memory for spatial and local cues: A comparison of a storing and a nonstoring species. *Anim. Learn. Behav.* 22:119–133.

Brogden, W. J. 1939. Sensory pre-conditioning. *J. Exp. Psych.* 25:323–332.

Brooks R., and A. N. Meltzoff. 2002. The importance of eyes: How infants interpret adult looking behavior. *Dev. Psych.* 38:958–966.

Buckley, A. 1883. *The Winners in Life's Race: or, The Great Backboned Family.* D. Appleton, New York.

Budge, E. A. W. 1969. *The Gods of the Egyptians: or, Studies in Egyptian Mythology.* Dover Publications, New York.

Bugnyar, T., and B. Heinrich. 2005. Ravens, *Corvus corax,* differentiate between knowledgeable and ignorant competitors. *Proc. Roy. Soc. Lond. B* 272: 1641–1646.

Bugnyar, T., and L. Huber. 1997. Push or pull: An experimental study on imitation in marmosets. *Anim. Behav.* 54:817–831.

Bugnyar, T., and K. Kotrschal. 2002. Observational learning and the raiding of food caches in ravens, *Corvus corax*: Is it "tactical deception"? *Anim. Behav.* 64:185–195.

Bugnyar, T., and K. Kotrschal. 2004. Leading a conspecific away from food in ravens, *Corvus corax. Anim. Cog.* 7:69–76.

Bulger, J. B. 1993. Dominance rank and access to estrous females in male savanna baboons. *Behaviour* 127:67–103.

Bulger, J. B., and W. J. Hamilton. 1987. Rank and density correlates of inclusive fitness measures in a natural chacma baboon (*Papio ursinus*) population. *Int. J. Primatol.* 8:635–650.

Burkart, J., and A. Heschl. In press. Perspective taking or behaviour reading? Understanding visual access in common marmosets (*Callithrix jacchus*). *Anim. Behav.*

Burling, R. 2005. *The Talking Ape.* Oxford University Press, Oxford.

Busse, C. 1982. Leopard and lion predation upon chacma baboons living in the Moremi Wildlife Reserve. *Botswana Notes Rec.* 12:15–21.

Butterworth, G., and N. Jarrett. 1991. What minds have in common space: Spatial mechanisms serving joint visual attention in infancy. *Brit. J. Dev. Psych.* 9:55–72.

Byrne, R., and A. Whiten, eds. 1988. *Machiavellian Intelligence: Social Expertise and the Evolution of Intellect in Monkeys, Apes, and Humans.* Oxford University Press, Oxford.

Cacioppo, J. T., Ernst, J. M., Burleson, M. H; McClintock, M. K., Malarkey, W. B., Hawkley, L. C., Kowalewski, R. B., Paulsen, A., Hobson, J. A., Hugdahl, K., Spiegel, D., and G. G. Bernston. 2000. Lonely traits and concomitant physiological processes: The MacArthur social neuroscience studies. *Int. J. Psychophysiol.* 35:143–154.

Caldwell, C. A., and A. Whiten. 2004. Testing for social learning and imitation in common marmosets, *Callithrix jacchus*, using an artificial fruit. *Anim. Cog.* 7:77–85.

Call, J., Agnetta, B., and M. Tomasello. 2000. Cues that chimpanzees do and do not use to find hidden objects. *Anim. Cog.* 3:23–34.

Call, J., Aureli, F., and F. B. M. de Waal. 2002. Postconflict third-party affiliation in stumptailed macaques. *Anim. Behav.* 63:209–216.

Calvin, W. H., and D. Bickerton. 2000. *Lingua Ex Machina: Reconciling Darwin with the Human Brain*. MIT Press, Cambridge, MA.

Capaldi, E. J. 1993. Animal number abilities: Implications for a hierarchical approach to instrumental learning. Pp. 191–209 in S. T. Boysen and E. J. Capaldi, eds., *The Development of Numerical Competence*. Lawrence Erlbaum Associates, Hillsdale, NJ.

Caplan, D. 1992. *Language: Structure, Processing, and Disorders*. MIT Press, Cambridge, MA.

Caramazza, A., and B. Z. Mahon. 2006. The organization of conceptual knowledge in the brain: The future's past and some future directions. *Cog. Neurosci.* 23:13–38.

Carey, S., and E. Spelke. 1996. Science and core knowledge. *Phil. Sci.* 63: 515–533.

Carpenter, M., Tomasello, M., and E. S. Savage-Rumbaugh. 1995. Joint attention and imitative learning in children, chimpanzees, and enculturated chimpanzees. *Soc. Dev.* 4:217–237.

Carter, C. S. 1998. Neuroendocrine perspectives on social attachment and love. *Psychoneuroendocrinology* 23:779–818.

Chance, M. R. A., and A. P. Mead. 1953. Social behavior and primate evolution. *Symp. Soc. Exp. Biol. Evol.* 7:395–439.

Chapais, B. 1988. Rank maintenance in female Japanese macaques: Experimental evidence for social dependency. *Behaviour* 104:41–59.

Chapais, B. 2001. Primate nepotism: What is the explanatory value of kin selection? *Int. J. Primatol.* 22:203–229.

Cheney, D. L. 1983. Extra-familial alliances among vervet monkeys. Pp. 278–286 in R. A. Hinde, ed., *Primate Social Relationships*. Blackwell Scientific, Oxford.

Cheney, D. L., and R. M. Seyfarth. 1980. Vocal recognition in free-ranging vervet monkeys. *Anim. Behav.* 28:362–367.

Cheney, D. L., and R. M. Seyfarth. 1985a. Social and non-social knowledge in vervet monkeys. *Phil. Trans. Roy. Soc. Lond. B* 308:187–201.

Cheney, D. L., and R. M. Seyfarth. 1985b. Vervet monkey alarm calls: Manipulation through shared information? *Behaviour* 93:150–166.

Cheney, D. L., and R. M. Seyfarth. 1986. The recognition of social alliances among vervet monkeys. *Anim. Behav.* 34:1722–1731.

Cheney, D. L., and R. M. Seyfarth. 1989. Reconciliation and redirected aggression in vervet monkeys. *Behaviour* 110:258–275.

Cheney, D. L., and R. M. Seyfarth. 1990. *How Monkeys See the World*. University of Chicago Press, Chicago.

Cheney, D. L., and R. M. Seyfarth. 1997. Reconciliatory grunts by dominant female baboons influence victims' behaviour. *Anim. Behav.* 54:409–418.

Cheney, D. L., and R. M. Seyfarth. 1998. Why monkeys don't have language. Pp. 175–219 in G. Petersen, ed., *The Tanner Lectures on Human Values*, 19. University of Utah Press, Salt Lake City.

Cheney, D. L., and R. M. Seyfarth. 1999. Recognition of other individuals' social relationships by female baboons. *Anim. Behav.* 58:67–75.

Cheney, D. L., and R. M. Seyfarth. 2005. Social complexity and the information acquired during eavesdropping by primates and other animals. Pp. 583–603 in P. Ҝ. McGregor, ed., *Animal Communication Networks*. Cambridge University Press, Cambridge.

Cheney, D. L., Seyfarth, R. M., and J. B. Silk. 1995a. The responses of female baboons to anomalous social interactions: Evidence for causal reasoning? *J. Comp. Psych.* 109:134–141.

Cheney, D. L., Seyfarth, R. M., and J. B. Silk. 1995b. The role of grunts in reconciling opponents and facilitating interactions among adult female baboons. *Anim. Behav.* 50:249–257.

Cheney, D. L., Seyfarth, R. M., and R. Palombit. 1996. The function and mechanisms underlying baboon "contact" barks. *Anim. Behav.* 52:507–518.

Cheney, D. L., Seyfarth, R. M., Fischer, J., Beehner, J. C., Bergman, T. J., Johnson, S. E., Kitchen, D. M., Palombit, R. A., Rendall, D., and J. B. Silk. 2004. Factors affecting reproduction and mortality among baboons in the Okavango Delta, Botswana. *Int. J. Primatol.* 25:401–428.

Chomsky, N. 1959. A review of B. F. Skinner's *Verbal behavior*. *Language* 35:26–58.

Chomsky, N. 1975. *Reflections on Language*. Pantheon, New York.

Chomsky, N. 1982. *Noam Chomsky on the Generative Enterprise: A Discussion with Riny Huybregts and Henk van Riemsdijk*. Foris, Dordrecht.

Church, R. M., and W. H. Meck. 1984. The numerical attribute of stimuli. Pp. 445–464 in H. L. Roitblat, T. G. Bever, and H. S. Terrace, eds., *Animal Cognition*. Lawrence Erlbaum Associates, Hillsdale, NJ.

Clark, A. P., and R. W. Wrangham. 1994. Chimpanzee arrival pant-hoots: Do they signify food or status? *Int. J. Primatol.* 15:185–205.

Clayton, N. S., Bussey, T. J., Emery, N. J., and A. Dickinson. 2003. Prometheus to Proust: The case for behavioural criteria for "mental time travel." *Trends Cog. Sci.* 7:436–437.

Clayton, N. S., and A. Dickinson. 1998. Episodic-like memory during cache recovery by scrub jays. *Nature* 395:272–278.

Clayton, N. S., Yu, K. S., and A. Dickinson. 2003. Interacting cache memories: Evidence of flexible memory use by scrub jays. *J. Exp. Psych. Anim. Behav. Proc.* 29:14–22.

Clutton-Brock, T. H., and S. D. Albon. 1979. The roaring of red deer and the evolution of honest advertisement. *Behaviour* 69:145–170.

Clutton-Brock, T. H., and P. Harvey. 1980. Primates, brains, and ecology. *J. Zool. Lond.* 190:309–323.

Cohen, Y. E., Hauser, M. D., and B. E. Russ. 2006. Spontaneous processing of abstract categorical information in the ventrolateral prefrontal cortex. *Biol. Lett.* 2:261–265.

Colwill, R. M., and R. A. Rescorla. 1985. Postconditioning devaluation of a reinforcer affects instrumental responding. *J. Exp. Psych. Anim. Behav. Proc.* 11:120–132.

Combes, S. L., and J. Altmann. 2001. Status change during adulthood: Life-history byproduct or kin-selection based on reproductive value? *Proc. Roy. Soc. Lond. B.* 268:1367–1373.

Connor, R. C., Smolker, R. A., and A. F. Richards. 1992. Dolphin alliances and coalitions. Pp. 415–443 in A. H. Harcourt and F. de Waal, eds., *Coalitions and Alliances in Humans and Other Animals*. Oxford University Press, Oxford.

Connor, R. C., Mann, J., Tyack, P. L., and H. Whitehead. 1998. Social evolution in toothed whales. *Trends Ecol. Evol.* 13:228–232.

Connor, R. C., Heithaus, R. M., and L. M. Barre. 1999. Superalliance of bottlenose dolphins. *Nature* 371:571–572.

Connor, R. C., Heithaus, R. M., and L. M. Barre. 2001. Complex social structure, alliance stability, and mating access in a bottlenose dolphin "super-alliance." *Proc. Roy. Soc. Lond. B* 268:263–267.

Corbey, R. H. A. 2005. *The Metaphysics of Apes: Negotiating the Animal-Human Boundary.* Cambridge University Press, Cambridge.

Courage, M. L., Edison, S. C., and M. L. Howe. 2004. Variability in the early development of visual self-recognition. *Inf. Behav. Dev..* 27:509–532.

Cowlishaw, G. 1994. Vulnerability to predation in baboon populations. *Behaviour* 131:293–304.

Crockford, C., and C. Boesch. 2003. Context-specific calls in wild chimpanzees, *Pan troglodytes verus*: Analysis of barks. *Anim. Behav.* 66:115–125.

Crockford, C., Herbinger, I., Vigiland, L., and C. Boesch. 2004. Wild chimpanzees have group-specific calls: A case for vocal learning? *Ethology* 110:221–243.

Crockford, C., Wittig, R. M., Seyfarth, R. M., and D. L. Cheney. 2007. Baboons eavesdrop to deduce mating opportunities. *Anim. Behav.*

Curtiss, S. 1977. *Genie: A Linguistic Study of a Modern-Day "Wild Child."* Academic Press, New York.

D'Amato, M., and M. Colombo. 1989. Serial learning with wild card items by monkeys (*Cebus apella*): Implications for knowledge of ordinal rank. *J. Comp. Psych.*, 103:252–261.

Dally, J. M., Emery, N. J., and N. S. Clayton. 2005. Cache protection strategies by western scrub-jays: implications for social cognition. *Anim. Behav.* 70:1251–1263.

Dally, J. M., Emery, N. J., and N. S. Clayton. 2006. Food-caching western scrub jays keep track of who was watching when. *Science* 312:1662–1665.

Damon, W., and D. Hart. 1982. The development of self-understanding from infancy through adolescence. *Child Dev.* 53:841–864.

Darwin, C. 1838a. *Notebook C.* In P. H. Barrett et al. (1987).

Darwin, C. 1838b. *Notebook M.* In P. H. Barrett et al. (1987).

Darwin, C. 1839. *Old and Useless Notes.* In P. H. Barrett et al. (1987).

Darwin, C. 1871/1981. *The Descent of Man, and Selection in Relation to Sex.* Princeton University Press, Princeton.

Dasser, V. 1988. A social concept in Java monkeys. *Anim. Behav.* 36:225–230.

David, R. 1998. *Handbook of Life in Ancient Egypt.* Oxford University Press, Oxford.

de Kort, S. R., Dickinson, A., and N. S. Clayton. 2005. Retrospective cognition by food-caching western scrub-jays. *Learn. Motiv.* 36:159–176.

de Waal, F. B. M., and F. Aureli. 1996. Consolation, reconciliation, and a possible cognitive difference between macaques and chimpanzees. Pp. 80–110 in A. E. Russon, K. A. Bard, and S. T. Parker, eds., *Reaching into Thought: The Minds of the Great Apes.* Cambridge University Press, Cambridge.

Deacon, T. 1997. *The Symbolic Species.* W. W. Norton, New York.

Deaner, R. O., and M. L. Platt. 2003. Reflexive social attention in monkeys and humans. *Curr. Biol.* 13:1609–1613.

Decety, J., and P. L. Jackson. 2004. The functional architecture of human empathy. *Behav. Cog. Neurosci. Rev.* 3:71–100.

Decety, J., and P. L. Jackson. 2006. A social neuroscience perspective on empathy. *Curr. Dir. Psych. Sci.* 15:54–58.

Decety, J., and J. A. Sommerville. 2003. Shared representations between self and other: A social cognitive neuroscience view. *Trends Cog. Sci.* 7:527–533.

Dehaene-Lambertz, G., Dehaene, S., and L. Hertz-Pannier. 2002. Functional-neuroimaging of speech perception in infants. *Science* 298:2013–2015.

Dennett, D. C. 1987a. Consciousness. Pp. 160–164 in R. L. Gregory, ed., *The Oxford Companion to the Mind.* Oxford University Press, Oxford.

Dennett, D. C. 1987b. *The Intentional Stance.* MIT/Bradford Books, Cambridge, MA.

Dennett, D. C. 1991. *Consciousness Explained.* Little, Brown, Boston.

Dickinson, A. 1980. *Contemporary Animal Learning Theory.* Cambridge University Press, Cambridge.

Drummey, A. B., and N. S. Newcombe. 2002. Developmental changes in source memory. *Dev. Sci.* 5:502–513.

du Plessis, P. n. d. Jack the signalman. In *EarthFooters' Writings from South Africa.* http://www.earthfoot.org/lit_zone/zasubndx.htm.

Dube, W. V., McIlvaine, W. J., Callahan, T. D., and L. T. Stoddard. 1993. The search for stimulus equivalence in nonverbal organisms. *Psych. Rec.* 43: 761–778.

Dunbar, R. 2000. Causal reasoning, mental rehearsal, and the evolution of primate cognition. Pp. 205–219 in C. Heyes and L. Huber, eds., *The Evolution of Cognition.* MIT Press, Cambridge, MA.

Dunbar, R. 2003. Why are apes so smart? Pp. 285–298 in P. M. Kappeler and M.

E. Pereira, eds., *Primate Life Histories and Socioecology.* University of Chicago Press, Chicago.

Ehmer, B., Reeve, H. K., and R. R. Hoy. 2001. Comparison of brain volumes between single and multiple foundresses in the paper wasp *Polistes dominulus*. *Brain Behav. Evol.* 57:161–168.

Eichenbaum, H., Fortin, N. J., Ergorul, C., Wright, S. P., and K. L. Agster. 2005. Episodic recollection in animals: "If it walks like a duck and quacks like a duck..." *Learn. Motiv.* 36: 190–207.

Eifuku, S., De Souza, W. C., Tamura, R., Nishijo, H., and T. Ono. 2004. Neuronal correlates of face identification in the monkey anterior temporal cortical areas. *J. Neurophys.* 91:358–371.

Elowson, A. M., and C. T. Snowdon. 1994. Pygmy marmosets, *Cebuella pygmaea*, modify vocal structure in response to changed social environment. *Anim. Behav.* 47: 1267–1277.

Emery, N. J. 2000. The eyes have it: The neuroethology, function and evolution of social gaze. *Neurosci. Biobehav. Rev.* 24:581–604

Emery, N. J., and N. S. Clayton. 2001. Effects of experience and social context on prospective caching strategies in scrub jays. *Nature* 414:443–446.

Emery, N. J., and D. I. Perrett. 2000. How can studies of the monkey brain help us understand "theory of mind" and autism in humans? Pp. 279–310 in S. Baron-Cohen, H. Tager-Flusberg, H., and D. Cohen, eds., *Understanding Other Minds: Perspectives from Developmental Cognitive Neuroscience*, 2nd ed. Oxford University Press, Oxford.

Emery, N. J., Dally, J. M., and N. S. Clayton. 2004. Western scrub-jays (*Aphelocoma californica*) use cognitive strategies to protect their caches from thieving conspecifics. *Anim. Cog.* 7:37–43.

Enard, W., Przeworski, M., Fisher, S. E., Lai, C. S. L., Wiebe, V., Kitano, T., Monaco, A. P., and S. Pääbo. 2002. Molecular evolution of FOXP2: A gene involved in speech and language. *Nature* 418:869–872.

Engh, A. L., Esch, K., Smale, L., and K. E. Holekamp. 2000. Mechanisms of maternal rank "inheritance" in the spotted hyaena, *Crocuta crocuta*. *Anim. Behav.* 60:323–332.

Engh, A. L., Siebert, E. R., Greenberg, D. A., and K. E. Holekamp. 2005. Patterns of alliance formation and postconflict aggression indicate spotted hyaenas recognize third-party relationships. *Anim. Behav.* 69:209–217.

Engh, A. E., Beehner, J. C., Bergman, T. J., Whitten, P. L., Hoffmeier, R. R., Seyfarth, R. M., and D. L. Cheney. 2006a. Behavioural and hormonal responses to predation in female chacma baboons (*Papio hamadryas ursinus*). *Proc. Roy. Soc. Lond. B* 273:707–712.

Engh, A. L., Beehner, J. C., Bergman, T. J., Whitten, P. L., Hoffmeier, R. R., Seyfarth, R. M., and D. L. Cheney. 2006b. Female hierarchy instability, male immigration, and infanticide increase glucocorticoid levels in female chacma baboons. *Anim. Behav.* 71:1227–1237.

Engh, A. E., Hoffmeier, R. R., Cheney, D. L., and R. M. Seyfarth. 2006c.

Who, me?: Can baboons infer the target of vocalisations? *Anim. Behav.* 71: 381–387.

Ergorul, C., and H. Eichenbaum. 2004. The hippocampus and memory for "what," "where," and when." *Learn. Mem.* 11:397–405.

Fagot, J., Wasserman, E. A., and M. Young. 2001. Discriminating the relation between relations: The role of entropy in abstract conceptualization by baboons (*Papio papio*) and humans (*Homo sapiens*). *J. Exp. Psych. Anim. Behav. Proc.* 27:316–328.

Fairbanks, L. A. 1980. Relationships among adult females in captive vervet monkeys: Testing a model of rank-related attractiveness. *Anim. Behav.* 28: 853–859.

Faithfull, M., and D. Dalton. 1994. *Faithfull: An Autobiography.* Little, Brown, Boston.

Fichtel, C., and P. M. Kappeler. 2002. Anti-predator behavior of group-living Malagasy primates: Mixed evidence for a referential alarm call system. *Behav. Ecol. Sociobiol.* 51:262–275.

Figley, C. R., ed. 1995. *Compassion Fatigue: Coping with Traumatic Stress Disorder in Those Who Treat the Traumatized.* Brunner/Mazel Psychological Stress Series, Philadelphia.

Fine, C., Lumsden, J., and J. R. Blair. 2001. Dissociation between "theory of mind" and executive functions in a patient with early left amygdala damage. *Brain* 124:287–298.

Fischer, J. 1998. Barbary macaques categorize shrill barks into two call types. *Anim. Behav.* 55:799–807.

Fischer, J. 2004. Emergence of individual recognition in young macaques. *Anim. Behav.* 67:655–661.

Fischer, J. 2006. Categorical perception. Pp. 248–251 in K. Brown, ed., *Encyclopedia of Language and Linguistics,* 2nd ed. Elsevier, Oxford.

Fischer, J., Cheney, D. L., and R. M. Seyfarth. 2000. Development of infant baboon responses to female graded variants of barks. *Proc. Roy. Soc. Lond. B* 267:2317–2321.

Fischer, J., Hammerschmidt, K., Seyfarth, R. M., and D. L. Cheney. 2001a. Acoustic features of female chacma baboon barks. *Ethology* 107:33–54.

Fischer, J., Metz, M., Cheney, D. L., and R. M. Seyfarth. 2001b. Baboon responses to graded bark variants. *Anim. Behav.* 61:925–931.

Fischer, J., Hammerschmidt, K., Cheney, D. L., and R. M. Seyfarth. 2002. Acoustic features of male baboon loud calls: Influences of context, age, and individuality. *J. Acoust. Soc. Am.* 111:1465–1474.

Fischer, J., Kitchen, D. M., Seyfarth, R. M., and D. L. Cheney. 2004a. Baboon loud calls advertise male quality: Acoustic features and their relation to rank, age, and exhaustion. *Behav. Ecol. Sociobiol.* 56:140–148.

Fischer, J., Call, J., and J. Kaminski. 2004b. A pluralistic account of word learning. *Trends Cog. Sci.* 8:481.

Fisher, C., and L. R. Gleitman. 2002. Language acquisition. Pp. 445–496 in H. F.

Pashler and C. R. Gallistel, eds., *Stevens Handbook of Experimental Psychology, Vol. 3: Learning and Motivation.* Wiley, New York.

Fitch, W. T., and M. D. Hauser. 1995. Vocal production in nonhuman primates: Acoustics, physiology, and functional constraints on "honest" advertisement. *Am. J. Primatol.* 37:191–220.

Fitch, W. T., Neubauer, J., and H. Herzel. 2002. Calls out of chaos: The adaptive significance of nonlinear phenomena in mammalian vocal production. *Anim. Behav.* 63:407–418

Flombaum, J. I., and L. R. Santos. 2005. Rhesus monkeys attribute perceptions to others. *Curr. Biol.* 15:447–452.

Fodor, J. A. 1975. *The Language of Thought.* Crowell Press, New York.

Fodor, J. A. 1986. Why paramecia don't have mental representations. Pp. 3–23 in P. A. French, T. J. Uehling, and H. K. Wettstein, eds., *Midwest Studies in Philosophy, X: Studies in the Philosophy of Mind.* University of Minnesota Press, Minneapolis.

Fogassi, L., Ferrari, P. F., Gesierich, B., Rozzi, S., Chersi, F., and G. Rizzolatti. 2005. Parietal lobe: From action organization to intention understanding. *Science* 308:662–667.

Foster, D. J., and M. A. Wilson. 2006. Reverse replay of behavioural sequences in hippocampal place cells during the awake state. *Nature* 440:680–683.

Fragaszy, D. M., Izar, P., Visalberghi, E., Ottoni, E. B., and M. Gomes De Oliveira. 2004. Wild capuchin monkeys (*Cebus libidinosus*) use anvils and stone pounding tools. *Am. J. Primatol.* 64:359–366.

Fragaszy, D. M., and E. Visalberghi. 1989. Social influences on the acquisition of tool-using behaviors in tufted capuchin monkeys (*Cebus apella*). *J. Comp. Psych.* 103:159–170.

Freedman, D. J., Riesenhuber, M., Poggio, T., and E. K. Miller. 2001. Categorical representation of visual stimuli in the primate prefrontal cortex. *Science* 291:312–316.

Frith, U., and C. D. Frith. 2003. Development and neurophysiology of mentalizing. *Phil. Trans. Roy. Soc. Lond. B* 358:459–473.

Fuster, J. M. 1997. *The Prefrontal Cortex.* Lippincott-Raven, Philadelphia.

Gacsi, M., Miklosi, A., Varga, O., Topal, J., and V. Csanyi. 2004. Are readers of our face readers of our mind?: Dogs (*Canis familiaris*) show situation-dependent recognition of human's attention. *Anim. Cog.* 7:144–153.

Gallagher, H. L., and C. D. Frith. 2003. Functional imaging of "theory of mind." *Trends Cog. Sci.* 7:77–83.

Gallup, G. G. 1970. Chimpanzees: Self-recognition. *Science* 167:86–87.

Garner, R. L. 1892. *The Speech of Monkeys.* Charles L. Webster and Co, New York.

Garnham, W. A., and T. Ruffman. 2001. Doesn't see, doesn't know: Is anticipatory looking really related to understanding or belief? *Dev. Sci.* 4:94–100.

Gautier, J. P., and A. Gautier. 1977. Communication in Old World monkeys.

Pp. 890–964 in T. Sebeok, ed., *How Animals Communicate*. Indiana University Press, Bloomington.

Gelman, R., and E. Spelke. 1981. The development of thoughts about animate and inanimate objects. Pp. 43–66 in J. H. Flavell and L. Ross, eds., *Social Cognitive Development*. Cambridge University Press, Cambridge.

Ghazanfar, A., and N. Logothetis. 2003. Facial expressions linked to monkey calls. *Nature* 423:937–938.

Ghazanfar, A. A., and L. R. Santos. 2004. Primate brains in the wild: The sensory bases for social interactions. *Nat. Rev. Neurosci.* 5:603–616.

Ghazanfar, A. A., Maier, J. X., Hoffman, K. L., and N. Logothetis. 2005. Multisensory integration of dynamic faces and voices in rhesus monkey auditory cortex. *J. Neurosci.* 25:5004–5012.

Gifford, G. W., Hauser, M. D., and Y. E. Cohen. 2003. Discrimination of functionally referential calls by laboratory-housed rhesus macaques: Implications for neuroethological studies. *Brain Behav. Evol.* 61:213–224.

Gifford, G. W., MacLean, K. A., Hauser, M. D., and Y. E. Cohen. 2005. The neurophysiology of functionally meaningful categories: Macaque ventrolateral prefrontal cortex plays a critical role in spontaneous categorization of species-specific vocalizations. *J. Cog. Neurosci.* 17:1471–1482.

Gil da Costa, R., Braun, A., Lopes, M., Hauser, M. D., Carson, R. E., Herskovitch, P., and A. Martin. 2004. Toward an evolutionary perspective on conceptual representation: Species-specific calls activate visual and affective processing systems in the macaque. *Proc. Nat. Acad. Sci. USA* 101:17516–17521.

Ginsberg, L. 1968. *The Legends of the Jews, Vol. 1*. Jewish Publishing Society of America, Philadelphia.

Giurfa, M., Zhang, S., Jennet, A., Menzel, R., and M. V. Srinivasan. 2001. The concepts of "sameness" and "difference" in an insect. *Nature* 410: 930–933.

Gleitman, L. R., and A. Papafragou. 2005. Language and thought. Pp. 633–661 in K. Holyoak and R. Morrison, eds., *Cambridge Handbook of Thinking and Reasoning*. Cambridge University Press, Cambridge.

Goldin-Meadow, S. 2003. *The Resilience of Language: What Gesture Creation in Deaf Children Can Tell Us about How All Children Learn Language*. Psychology Press, New York.

Golinkoff, R. A. 1986. "I beg your pardon?": The preverbal negotiation of failed messages. *J. Child Lang.* 13:455–476.

Goodall, J. van Lawick. 1968. The behaviour of free-living chimpanzees in the Gombe Stream Reserve. *Anim. Behav. Monographs* 1:165–311.

Goodall, J. 1986. *The Chimpanzees of Gombe: Patterns of Behavior*. Harvard University Press, Cambridge, MA.

Gopnik, A., and P. Graf. 1988. Knowing how you know: Young children's ability to identify and remember the sources of their beliefs. *Child Dev.* 59:1366–1371.

Gopnik, A., and A. N. Meltzoff. 1994. Minds, bodies, and persons: Young children's understanding of the self and others as reflected in imitation and theory of mind research. Pp. 166–186 in S. T. Parker, R. W. Mitchell, and M. Boccia, eds., *Self Awareness in Animals and Humans*. Cambridge University Press, Cambridge.

Goymann, W., and J. C. Wingfield. 2004. Allostatic load, social status, and stress hormones: The costs of social status matter. *Anim. Behav.* 67:591–602.

Grandin, T., and C. Johnson. 2005. *Animals in Translation: Using the Mysteries of Autism to Decode Animal Behavior.* Scribner, New York.

Grice, H. P. 1957. Meaning. *Phil. Rev.* 66:377–388.

Griffiths, D. P., Dickinson, A., and N. S. Clayton. 1999. Declarative and episodic memory: What can animals remember about their past? *Trends Cog. Sci.* 3:74–80.

Gros-Louis, J. 2004. The function of food-associated calls in white-faced capuchin monkeys, *Cebus capucinus*, from the perspective of the signaler. *Anim. Behav.* 67:431–440.

Gust, D. A., Gordon, T. P., Hambright, M. K., and M. E. Wilson. 1993. Relationship between social factors and pituitary–adrenocortical activity in female rhesus monkeys (*Macaca mulatta*). *Horm. Behav.* 27:318–331.

Gust, D. A., Gordon, T. P., Brodie, A. R., and H. M. McClure. 1994. Effect of a preferred companion in modulating stress in adult female rhesus monkeys. *Physiol. Behav.* 55:681–684.

Gyger, M., Karakashian, S. J., and P. Marler. 1986. Avian alarm-calling: Is there an audience effect? *Anim. Behav.* 34:1570–1572.

Hage, S., Jurgens, U., and G. Ehret. 2006. Audio-vocal interaction in the pontine brainstem during self-initiated vocalization in the squirrel monkey. *Eur. J. Neurosci.* 23:3297–3308.

Hamilton, W. D. 1964. The genetical evolution of social behaviour. *J. Theor. Biol.* 7:1–51.

Hamilton, W. J., Buskirk, R. E., and W. H. Buskirk. 1975. Defensive stoning by baboons. *Nature* 256:488–489.

Hamilton, W. J., Buskirk, R. E., and W. H. Buskirk. 1976. Defense of space and resources by chacma (*Papio ursinus*) baboon troops in an African desert and swamp. *Ecology* 57:1264–1272.

Hammerschmidt, K., and D. Todt. 1995. Individual differences in vocalisations in young Barbary macaques (*Macaca sylvanus*): A multi-parametric analysis to identify critical cues in acoustic signalling. *Behaviour* 132:381–399.

Hammerschmidt, K., and U. Jurgens. In press. Acoustical correlates of affective prosody. *Voice*.

Hampton, R. R. 2001. Rhesus monkeys know when they remember. *Proc. Nat. Acad. Sci. USA* 98:5359–5362.

Hampton, R. R. 2005. Can rhesus monkeys discriminate between remembering and forgetting? Pp. 272–295 in H. S. Terrace and J. Metcalfe, eds., *The Missing Link in Cognition: Origins of the Self-Reflective Consciousness*. Oxford University Press, New York.

Hare, B., and M. Tomasello. 1999. Domestic dogs (*Canis familiaris*) use human and conspecific social cues to locate hidden food. *J. Comp. Psych.* 113:1–5.

Hare, B., and M. Tomasello. 2004. Chimpanzees are more skilful in competitive than in cooperative cognitive tasks. *Anim. Behav.* 68:571–581.

Hare, B., Call, J., and M. Tomasello. 1998. Communication of food location between human and dog (*Canis familiaris*). *Evol. Comm.* 2:137–159.

Hare, B., Call, J., Agnetta, B., and M. Tomasello. 2000. Chimpanzees know what conspecifics do and do not see. *Anim. Behav.* 59:771–785.

Hare, B., Call, J., and M. Tomasello. 2001. Do chimpanzees know what conspecifics know? *Anim. Behav.* 61:139–151.

Hare, B., Brown, M., Williamson, C., and M. Tomasello. 2002. The domestication of social cognition in dogs. *Science* 298:1636–1639.

Harlow, H. F. 1949. The formation of learning sets. *Psych. Rev.* 56:51–65.

Harraway, D. 1989. *Primate Visions: Gender, Race, and Nature in the World of Modern Science.* Routledge, New York.

Hauser, M. D. 1988. How infant vervet monkeys learn to recognize starling alarm calls: The role of experience. *Behaviour* 105:187–201.

Hauser, M. D. 1992a. Articulatory and social factors influence the acoustic structure of rhesus monkey vocalizations: A learned mode of production? *J. Acoust. Soc.* 91:2175–2179.

Hauser, M. D. 1992b. Costs of deception: Cheaters are punished in rhesus monkeys. *Proc. Nat. Acad. Sci. USA* 89:12137–12139.

Hauser, M. D. 1996. *The Evolution of Communication.* MIT Press, Cambridge, MA.

Hauser, M. D., Chomsky, N., and W. Fitch. 2002. The faculty of language: What is it, who has it, and how did it evolve? *Science* 298:1565–1568.

Hefner, H. E., and R. S. Hefner. 1984. Temporal lobe lesions and perception of species-specific vocalizations by macaques. *Science* 226:75–76.

Hemelrijk, C. K. 1994. Support for being groomed in long-tailed macaques, *Macaca fascicularis.* *Anim. Behav.* 48:479–481.

Henkin, L. 1940. *Darwinism in the English Novel, 1860–1910.* Corporate Press, New York.

Herman, L. M, Pack, A. A, and P. Morrel-Samuels. 1993a. Representational and conceptual skills of dolphins. Pp. 403–442 in H. L. Roitblat, L. M. Herman, and P. E. Nachtigall, eds., *Comparative Cognition and Neuroscience.* Lawrence Erlbaum Associates, Hillsdale, NJ.

Herman, L. M., Kuczaj, S. A., and M. D. Holder. 1993b. Responses to anomalous gestural sequences by a language-trained dolphin: Evidence for processing of semantic relations and syntactic information. *J. Exp. Psych. Gen.* 122:184–194.

Heyes, C. M. 1994. Social cognition in primates. Pp. 281–305 in N. J. Macintosh, ed., *Animal Learning and Cognition.* Academic Press, New York.

Hihara, S., Yamada, H., Iriki, A., and K. Okanoya. 2003. Spontaneous vocal differentiation of coo-calls for tools and food in Japanese monkeys. *Neurosci. Res.* 45:383–389.

Hodge, J. 2003. The notebook programmes and projects of Darwin's London years. Pp. 40–68 in J. Hodge and G. Radick, eds., *The Cambridge Companion to Darwin*. Cambridge University Press, Cambridge.

Hoesch, von W. 1961. Uber Ziegen hütende Bärenpaviane (*Papio ursinus rua-cana*). *Z. Tierpsychol.* 18:297–301.

Hogrefe, G. J., Wimmer, H., and J. Perner. 1986. Ignorance versus false belief: A developmental lag in attribution of epistemic states. *Child Dev.* 57:567–582.

Holekamp, K. E., Boydston, E. E., Szykman, M., Graham, I., Nutt, K. J., Birch, S., Piskiel, A., and M. Singh. 1999. Vocal recognition in the spotted hyena and its possible implications regarding the evolution of intelligence. *Anim. Behav.* 58:383–395.

Hrdy, S. B.1977. *The Langurs of Abu*. Harvard University Press, Cambridge, MA.

Humphrey, N. K. 1976. The social function of intellect. Pp. 303–318 in P. Bateson and R. A. Hinde, eds., *Growing Points in Ethology*. Cambridge University Press, Cambridge.

Humphrey, N. K. 1986. *The Inner Eye*. Faber & Faber, London.

Humphreys, G. W., and M. J. Riddoch. 2006. Features, objects, action: The cognitive neuropsychology of visual object processing. *Cog. Neuropsych.* 23:156–183.

Hunt, G. R., and R. D. Gray. 2004. Direct observations of pandanus-tool manufacture and use by a New Caledonian crow (*Corvus moneduloides*). *Anim. Cog.* 7:114–120.

Hurford, J. R. 1998. Introduction: The emergence of syntax. Pp. 299–304 in J. R. Hurford, M. Studdert-Kennedy, and C. Knight, eds., *Approaches to the Evolution of Language*. Cambridge University Press, Cambridge.

Hurford, J. R. 2003. The neural basis of predicate-argument structure. *Behav. Brain Sci.* 26:261–316.

Irwin, M. R., Risch, S. C., Daniels, M., Bloom, E., and H. Weiner. 1987. Plasma-cortisol and immune function in bereavement. *Psychosom. Med.* 49:210–211.

Jackendoff, R. 1987. *Consciousness and the Computational Mind*. Basic Books, New York.

Jackendoff, R. 1990. What would a theory of language evolution have to look like? *Behav. Brain Sci.* 13:737–738.

Jackendoff, R. 1994. *Patterns in the Mind*. Basic Books, New York.

Jackendoff, R. 1999. Possible stages in the evolution of the language capacity. *Trends Cog. Sci.* 3:272–279.

Jackendoff, R. 2002. *Foundations of Language*. Oxford University Press, Oxford.

James, William 1872/1982. *Psychology: Briefer Course*. Harvard University Press, Cambridge, MA.

Janson, C. H., and C. P. van Schaik. 2000. The behavioral ecology of infanticide by males. Pp. 469–494 in C. P. van Schaik and C. H. Janson, eds., *Infanticide by Males and Its Implications*. Cambridge University Press, Cambridge.

Janson, H. W. 1952. *Apes and Ape Lore in the Middle Ages and the Renaissance.* Warburg Institute, London.

Janssen, R., and J. Janssen. 1989. *Egyptian Household Animals.* Shire Publishing, Aylesbury, UK.

Jellema, T., Baker, C. I., Wicker B., and D. I. Perrett. 2000. Neural representation for the perception of the intentionality of actions. *Brain Cogn.* 44:280–302.

Jensen, K., Hare, B., Call, J., and M. Tomasello. 2006. What's in it for me?: Self-regard precludes altruism and spite in chimpanzees. *Proc. Roy. Soc. Lond. B* 273:1013–1021.

Jobson, R. 1623. *The Golden Trade, or A discovery of the river Gambra, and the Golden Trade of the Aethiopians* (Charles G. Kingsley, ed). Teignmouth, London.

Johansson, S. 2005. *Origins of Language: Constraints on Hypotheses.* John Benjamins Publishing, Philadelphia.

Johnson-Laird, P. N. 1983. *Mental Modes.* Cambridge University Press, Cambridge.

Jolly, A. 1966. Lemur social behavior and primate intelligence. *Science* 153: 501–506.

Judge, P. 1982. Redirection of aggression based on kinship in a captive group of pigtail macaques. *Int. J. Primatol.* 3:301.

Judson, H. F. 1979. *The Eighth Day of Creation.* Simon and Schuster, New York.

Julian, G. E., and W. Gronenberg. 2002. Reduction of brain volume correlates with behavioral changes in queen ants. *Brain Behav. Evol.* 60:152–164.

Jurgens, U. 1995. Neuronal control of nonhuman and human primates. Pp. 199–207 in E. Zimmerman, J. D. Newman, and U. Jurgens, eds., *Current Topics in Primate Vocal Communication.* Plenum Press, New York.

Kamil, A. C., Balda, R. P., and D. J. Olson. 1994. Performance of four seed-caching corvid species in the radial-arm maze analog. *J. Comp. Psych.* 108: 385–393.

Kaminski, J., Call, J., and J. Fischer. 2004. Word learning in a domestic dog: Evidence for "fast mapping." *Science* 304:1682–1683.

Kaplan, H., Hill, K., Lancaster, J., and A. M. Hurtado. 2000. A theory of human life history evolution: Diet, intelligence, and longevity. *Evol. Anthr.* 9:156–185.

Kappeler, P. M., and C. P. van Schaik. 1992. Methodological and evolutionary aspects of reconciliation among primates. *Ethology* 92:51–69.

Karakashian, S. J., Gyger, M., and P. Marler.1988. Audience effects on alarm-calling in chickens. *J. Comp. Psych.* 102:129–135.

Karin-D'Arcy, M. R., and D. J. Povinelli. 2002. Do chimpanzees know what others see?: A closer look. *Int. J. Comp. Psych.* 15:21–54.

Kawai, M. 1958. On the system of social ranks in a natural group of Japanese monkeys. *Primates* 1:11–48.

Kawamura, S. 1958. Matriarchal social order in the Minoo-B troop: A study of the rank system of Japanese monkeys. *Primates* 1:149–156.

Kawashima, S., Sugiura, M., Kato, T., Nakamura, A., Hatano, K., Ito, K., Fukuda,

H., Kojima, S., and K. Nakamura. 1999. The human amygdala plays an important role in gaze monitoring: A PET study. *Brain* 122:779–783.

Kellman, P. J., and E. S. Spelke. 1983. Perception of partially occluded objects in infancy. *Cog. Psych.* 15:483–524.

Kendler, K. S., Myers, J., and C. A. Prescott. 2005. Sex differences in the relationship between social support and risk for major depression: A longitudinal study of opposite sex twin pairs. *Am. J. Psych.* 162:250–256.

Kinsbourne, M. 2005. A continuum of self-consciousness that emerges in phylogeny and ontogeny. Pp. 142–156 in H. S. Terrace and J. Metcalfe, eds., *The Missing Link in Cognition: Origins of the Self-Reflective Consciousness.* Oxford University Press, New York.

Kirby, S. 1998. Fitness and the selective adaptation of language. Pp. 359–383 in J. R. Hurford, M. Studdert-Kennedy, and C. Knight, eds., *Approaches to the Evolution of Language.* Cambridge University Press, Cambridge.

Kitchen, D. M., Cheney, D. L., and R. M. Seyfarth. 2003a. Female baboons' responses to male loud calls. *Ethology* 109:401–412.

Kitchen, D. M., Seyfarth, R. M., Fischer, J., and Cheney, D. L. 2003b. Loud calls as an indicator of dominance in male baboons (*Papio cynocephalus ursinus*). *Behav. Ecol. Sociobiol.* 53:374–384.

Kitchen, D. M., Cheney, D. L., and R. M. Seyfarth. 2004. Factors mediating intergroup encounters in savannah baboons (*Papio cynocephalus ursinus*). *Behaviour* 141:197–218.

Kitchen, D. M., Cheney, D. L., and R. M. Seyfarth. 2005a. Contextual factors mediating contests between male chacma baboons in Botswana: The effect of food, friends, females. *International Journal of Primatology* 26: 105–125.

Kitchen, D. M., Cheney, D. L., and R. M. Seyfarth. 2005b. Male chacma baboons (*Papio hamadryas ursinus*) discriminate loud call contests between rivals of different relative ranks. *Anim. Cog.* 8:1–6.

Knight, C., Studdert-Kennedy, M., and J. R. Hurford. 2000. Language: A Darwinian adaptation? Pp. 1–15 in C. Knight, M. Studdert-Kennedy and J. R. Hurford, eds., *The Evolutionary Emergence of Language.* Cambridge University Press, Cambridge.

Kohler, W. 1925/1959. *The Mentality of Apes.* Vintage Books, New York.

Kummer, H. 1968. *Social Organization of Hamadryas Baboons.* University of Chicago Press, Chicago.

Kummer, H., Gotz, W., and W. Angst. 1974. Triadic differentiation: An inhibitory process protecting pair bonds in baboons. *Behaviour* 49:62–87.

Kummer, H., Anzenberger, G., and C. K. Hemelrijk. 1996. Hiding and perspective taking in long-tailed macaques (*Macaca fascicularis*). *J. Comp. Psych.* 110:97–102.

Landau, B., and L. R. Gleitman. 1985. *Language and Experience: Evidence from the Blind Child.* Harvard University Press, Cambridge, MA.

Langen, T. A. 1996. Social learning of a novel foraging skill by white-throated magpie jays (*Calocitta formosa*, Corvidae): A field experiment. *Ethology* 102:157–166.

Le Prell, G. C., Hauser, M. D., and D. B. Moody. 2002. Discrete or graded variation within rhesus monkey screams?: Psychophysical experiments on classification. *Anim. Behav.* 63:47–62.

Lefebvre, L., Palameta, B., and K. K. Hatch. 1996. Is group-living associated with social learning?: A comparative test of a gregarious and a territorial columbid. *Behaviour* 133:241–261.

Leonard, C. M., Rolls, E. T., Wilson, F. A. W., and G. C. Baylis. 1985. Neurons in the amygdala of the monkey with responses selective for faces. *Behav. Brain Res.* 15:159–176.

Linnaeus, C. 1735. *Systema Naturae.*

Livingstone, D. 1858. *Missionary Travels and Researches in South Africa.* Harper and Brothers, New York.

Logothetis, N., Pauls, J., Augath, M., Trinath, T., and A. Oltermann. 2001. Neurophysiological investigation of the basis of the fMRI signal. *Nature* 412: 150–157.

Lonsdorf, E. V. 2006. The role of the mother in the acquisition of tool-use skills in wild chimpanzees. *Anim. Cog.* 9:36–46.

Lorincz, E. N., Jellema, T., Gómez, J.-C., Barraclough, N., Xiao, D., and D. I. Perrett. 2005. Do monkeys understand actions and minds of others?: Studies of single cells and eye movements. Pp.189–210 in S. Dehaene, J.-R. Duhamel, M. D. Hauser, and G. Rizzolatti, eds., *From Monkey Brain to Human Brain.* MIT Press, Cambridge, MA.

Macuda, T., and W. A. Roberts. 1995. Further evidence for hierarchical chunking in rat spatial memory. *J. Exp. Psych. Anim. Behav. Proc.* 21: 20–32.

Malle, B. F., Moses, L. J., and D. A. Baldwin, eds. 2001. *Intentions and Intentionality: Foundations of Social Cognition.* MIT Press, Cambridge, MA.

Manser, M. B. 2001. The acoustic structure of suricates' alarm calls varies with predator type and the level of response urgency. *Proc. Roy. Soc. London Ser. B* 268: 2315–2324.

Manser, M. B., Bell, M. B., and L. Fletcher. 2001a. The information that receivers extract from alarm calls in suricates. *Proc. R. Soc. London Ser. B* 268: 2485–2491.

Manser, M., Seyfarth, R. M., and D. L. Cheney. 2001b. Suricate alarm calls signal predator class and urgency. *Trends Cog. Sci.* 6: 55–57.

Manson, J. H., Navarette, C. D., Silk, J. B., and S. Perry. 2004. Time-matched grooming in female primates?: New analyses from two species. *Anim. Behav.* 67: 493–500.

Marais, E. 1922/1969. *The Soul of the Ape.* Penguin, Harmondsworth.

Marais, E. 1939. *My Friends the Baboons.* Human and Rousseau, Cape Town.

Marino, L. 1998. A comparison of encephalization between odontocete cetaceans and anthropoid primates. *Brain Behav. Evol.* 51:230–238.

Markman, E. M., and M. Abelev. 2004. Word learning in dogs? *Trends Cog. Sci.* 8:479–481.

Marler, P. 1976. An ethological theory of the origin of vocal learning. Pp. 386–395 in S. R. Harnad, H. D. Steklis, and J. Lancaster, eds., *Origins and Evolution of Language and Speech.* New York Academy of Science, New York.

Marler, P., and S. Peters. 1988. The role of song phonology and syntax in vocal learning preferences in the song sparrow, *Melospiza melodia. Ethology* 77:125–149.

Marler, P., and S. Peters. 1989. Species differences in auditory responsiveness in early vocal learning. Pp. 243–273 in R. J. Dooling and S. Hulse, eds., *The Comparative Psychology of Audition: Perceiving Complex Sounds.* Lawrence Erlbaum, Hillsdale, NJ.

Marmot, M. G. 2004. *The Status Syndrome: How Social Standing Affects Our Health and Longevity.* Henry Holt, New York.

Martin, A. 1992. Semantic knowledge in patients with Alzheimer's disease: Evidence for degraded representations. Pp. 119–134 in L. Backman, ed., *Memory Functions in Dementia.* Elsevier/North Holland, Amsterdam.

Martin, A. 1998. Organization of semantic knowledge and the origin of words in the brain. Pp. 69–88 in N. G. Jablonski and L. C. Aiello, eds., *The Origin and Diversification of Language.* California Academy of Sciences, San Francisco.

Martin, R. D. 1983. *Human Brain Evolution in an Ecological Context.* American Museum of Natural History, New York.

Martin, R. D. 1990. *Primate Origins and Evolution: A Phylogenetic Reconstruction.* Princeton University Press, Princeton.

Masserman, J. H., Wechkin, S., and W. Terris. 1964. "Altruistic" behavior in rhesus monkeys. *Am. J. Psychiatr.* 121:584–585.

Maynard Smith, J. 1982. *Evolution and the Theory of Games.* Cambridge University Press, Cambridge.

Maynard Smith, J., and D. G. C. Harper. 1995. Animal signals: Models and terminology. *J. Theor. Biol.* 177:305–311.

McBeath, M. K., Shaffer, D. M., and M. K. Kaiser. 1995. How baseball outfielders determine where to run to catch fly balls. *Science* 268:569–573.

McCleery, J. M., Bhagwagar, Z., Smith, K. A., Goodwin, G. M., and P. J. Cowen. 2000. Modelling a loss event: Effect of imagined bereavement on the hypothalamic-pituitary-adrenal axis. *Psych. Med.* 30:219–223.

McDermott, W. C. 1938. *The Ape in Antiquity.* Johns Hopkins University Press, Baltimore.

McGregor, P., ed. 2005. *Animal Communication Networks.* Cambridge University Press, Cambridge.

Mead, G. H. 1964. *Selected Writings* (A. J. Reck, ed.). University of Chicago Press, Chicago.

Mendelsohn, J., and S. el Obeid. 2004. *Okavango River: The Flow of a Lifeline.* Struik Publishers, Cape Town.

Mennill, D. J., Ratcliffe, L. M., and P. T. Boag. 2002. Female eavesdropping on male song contests in songbirds. *Science* 296:873.

Mercado, E., Killebrew, D. A., Pack, A. A., Macha, I., and L. M. Herman. 2000. Generalization of "same-different" classification abilities in bottlenosed dolphins. *Behav. Proc.* 50:79–94.

Metcalfe, J., and H. Kober. 2005. Self-reflective consciousness and the project-able self. Pp. 57–83 in H. S. Terrace and J. Metcalfe, eds., *The Missing Link in Cognition: Origins of the Self-Reflective Consciousness.* Oxford University Press, New York.

Miklosi, A., and J. Topal. 2004. What can dogs teach us? *Anim. Behav.* 67:995–1004.

Mill, J. S. 1840. Review of the works of Samuel Taylor Coleridge. *Lond. Westminster Rev.* 33:257–302.

Mill, J. S. 1869. The subjection of women. Pp. 125–242 in A. S. Rossi, ed., *Essays on Sex Equality by John Stuart Mill and Harriet Taylor Mill.* University of Chicago Press, Chicago.

Miller, E. K., Freedman, D. J., and J. D. Wallis. 2002. The prefrontal cortex: Categories, concepts, and cognition. *Phil. Trans. Roy. Soc. Lond. B* 357:1123–1136.

Miller, G. A. 1956. The magical number seven, plus or minus two: Some limits on our capacity for processing information. *Psych. Rev.* 63: 81–97.

Milton, K. 1988. Foraging behavior and the evolution of primate intelligence. Pp. 285–306 in R. W. Byrne and A. Whiten, eds., *Machiavellian Intelligence: Social Expertise and the Evolution of Intellect in Monkeys, Apes, and Humans.* Oxford University Press, Oxford.

Mitani, J. 1993. Contexts and social correlates of long-distance calling by male chimpanzees. *Anim. Behav.* 45:735–746.

Mitani, J. C., and K. L. Brandt. 1994. Social factors influence the acoustic variability in the long-distance calls of male chimpanzees. *Ethology* 96:233–252.

Mitchell, J. P., Banaji, M. R., and C. N. Macrae. 2005. The link between social cognition and self-referential thought in the medial prefrontal cortex. *J. Cog. Neurosci.* 17:1306–1315.

Mulcahy, N. J., and J. Call. 2006. Apes save tools for future use. *Science* 312:1038–1040.

Naguib, M., Fichtel, C., and D. Todt. 1999. Nightingales respond more strongly to vocal leaders of simulated dyadic interactions. *Proc. Roy. Soc. Lond. B* 266:537–542.

Nakahara, K., and T. Miyashita. 2005. Understanding intentions: Through the looking glass. *Science* 308:6444–6445.

Neiworth, J. J., Burman, M. A., Basile, B. M., and M. T. Lickteig. 2002. Use of experimenter-given cues in visual co-orienting and in an object-choice task by a New World monkey species, cotton top tamarins (*Saguinus oedipus*). *J. Comp. Psych.* 116:3–11.

Nelson, K. 2005. Emerging levels of consciousness in early human development. Pp. 116–141 in H. S. Terrace and J. Metcalfe, eds., *The Missing Link in Cog-*

nition: Origins of the Self-Reflective Consciousness. Oxford University Press, New York.

Newmeyer, F. 1991. Functional explanations in linguistics and the origins of language. *Lang. Comm.* 11:3–28.

Newmeyer, F. 2003. Grammar is grammar and usage is usage. *Language* 79: 682–707.

Nishida, T. 1987. Local traditions and cultural transmission. Pp. 462–474 in B. B. Smuts, D. L. Cheney, R. M. Seyfarth, R. W. Wrangham, and T. T. Struhsaker, eds., *Primate Societies.* University of Chicago Press, Chicago.

Oden, D. L., Thompson, R. K. R., and D. Premack. 1988. Spontaneous transfer of matching by infant chimpanzees (*Pan troglodytes*). *J. Exp. Psych. Anim. Behav. Proc.* 14:140–145.

Ohnuki-Tierney, E. 1987. *The Monkey as Mirror: Symbolic Transformations in Japanese History and Ritual.* Princeton University Press, Princeton.

Oliveira, R. F., McGregor, P. K., and C. Latruffe. 1998. Know thine enemy: Fighting fish gather information from observing conspecific interactions. *Proc. Roy. Soc. Lond.* B 265:1045–1049.

Olson, D. J., Kamil, A. C., Balda, R. P., and P. J. Nims. 1995. Performance of four seed-caching Corvid species in operant tests of nonspatial and spatial memory. *J. Comp. Psych.* 109:173–181.

O'Neill, D. K. 1996. Two-year-old children's sensitivity to a parent's knowledge state when making requests. *Child Dev.* 67:659–677.

Onishi, K. H., and R. Baillargeon. 2005. Do 15-month-old infants understand false beliefs? *Science* 308:255–258.

Owings, D. H., and D. Hennessy. 1984. The importance of variation in sciurid visual and vocal communication. Pp. 202–247 in J. O. Murie and G. R. Michener, eds., *Biology of Ground-Dwelling Squirrels: Annual Cycles, Behavioral Ecology, and Sociality.* University of Nebraska Press, Lincoln.

Owings, D. H., and E. S. Morton. 1998. *Animal Vocal Communication: A New Approach.* Cambridge University Press, Cambridge.

Owren, M. J. 2000. Standing evolution on its head: The uneasy role of evolutionary theory in comparative cognition and communication. *Rev. Anthr.* 29: 55–69.

Owren, M. J., and D. Rendall. 1997. An affect-conditioning model of nonhuman primate vocal signaling. Pp. 299–346 in M. D. Beecher, D. H. Owings, and N. S. Thompson, eds., *Perspectives in Ethology, Vol. 12.* Plenum Press, New York.

Owren, M. J., Dieter, J. A., Seyfarth, R. M., and D. L. Cheney. 1992. "Food" calls produced by adult female rhesus and Japanese macaques, their normally raised offspring, and offspring cross-fostered between species. *Behaviour* 120:218–231.

Owren, M. J., Dieter, J. A., Seyfarth, R. M., and D. L. Cheney. 1993. Vocalizations of rhesus and Japanese macaques cross-fostered between species show evidence of only limited modification. *Dev. Psychobiol.* 26:389–406.

Owren, M. J., Seyfarth, R. M., and D. L. Cheney. 1997. The acoustic features of vowel-like grunt calls in chacma baboons (*Papio cynocephalus ursinus*): Implications for production processes and functions. *J. Acoust. Soc. Am.* 101:2951–2963.

Päabo, S. 2003. An ape perspective on human uniqueness. Paper presented at the 8th Congress of the German Primate Society, Leipzig, Germany, October 1–4, 2003.

Palombit, R. A., Seyfarth, R. M., and D. L. Cheney. 1997. The adaptive value of friendships to female baboons: Experimental and observational evidence. *Anim. Behav.* 54:599–614.

Palombit, R. A., Cheney, D. L., Fischer, J., Johnson, S., Rendall, D., Seyfarth, R. M., and J. B. Silk. 2000. Male infanticide and defense of infants in wild chacma baboons. Pp. 123–152 in C. P. van Schaik and C. H. Janson, eds., *Infanticide by Males and Its Implications*. Cambridge University Press, Cambridge.

Palombit, R. A., Cheney, D. L., and R. M. Seyfarth. 2001. Female-female competition for male "friends" in wild chacma baboons (*Papio cynocephalus ursinus*). *Anim. Behav.* 61:1159–1171.

Panksepp, J. 1998. *Affective Neuroscience: The Foundations of Human and Animal Emotions*. Oxford University Press, New York.

Passingham, R. E. 1982. *The Human Primate*. W. H. Freeman, San Francisco.

Paz-y-Miño, G., Bond, A. B., Kamil, A. C., and R. P. Balda. 2004. Pinyon jays use transitive inference to predict social dominance. *Nature* 430: 778–782.

Peacock, T. L. 1817/1896. *Melincourt*. Macmillan, London.

Peake, T. M., Terry, A. M. R., McGregor, P. K., and T. Dabelsteen. 2001. Male great tits eavesdrop on simulated male-to-male vocal interactions. *Proc. Roy. Soc. Lond. B* 268:1183–1187.

Peake, T. M., Terry, A. M. R., McGregor, P. K., and T. Dabelsteen. 2002. Do great tits assess rivals by combining direct experience with information gathered by eavesdropping? *Proc. Roy. Soc. Lond. B* 269:1925–1929.

Pears, D. F. 1987. Wittgenstein's philosophy of language. Pp. 811–813 in R. L. Gregory, ed., *The Oxford Companion to the Mind*. Oxford University Press, Oxford.

Pelphrey, K. A., Viola, R. J., and G. McCarthy. 2004. When strangers pass: Processing of mutual and averted social gaze in the superior temporal sulcus. *Psych. Sci.* 15:598–603.

Peña, M., Maki, A., Kovacic, D., Dehaene-Lambertz, G., Koizumi, H., Bouquet, F., and J. Mehler. 2003. Sounds and silence: An optical topography study of language recognition at birth. *Proc. Nat. Acad. Sci. USA* 100:11702–11705.

Pepperberg, I. M. 1992. Proficient performance of a conjunctive, recursive task by an African gray parrot (*Psittacus erithacus*). *J. Comp. Psych.* 106:295–305.

Pepperberg, I. M. 1994. Numerical competence in an African gray parrot (*Psittacus erithacus*). *J. Comp. Psych.* 108:36–44.

Perez-Barberia, F. J., and I. J. Gordon. 2005. Gregariousness increases brain size in ungulates. *Oecologia* 145:41–52.

Perner, J., and T. Ruffman. 2005. Infants' insight into the mind: How deep? *Science* 308:214–216.

Perrett, D. I., and N. J. Emery. 1994. Understanding the intentions of others from visual signals: Neurophysiological evidence. *Curr. Psych. Cog.* 13:683–694.

Perrett, D. I., Harries, M. H, Mistlin, A. J., and J. K. Hietanen. 1990. Social signals analyzed at the single cell level: Someone is looking at me, something touched me, something moved. *Int. J. Comp. Psych.* 4:25–55.

Perrett, D. I., Hietanen, J. K., Oram, M. W., and P. J. Benson. 1992. Organization and function of cells responsive to faces in the temporal cortex. *Phil. Trans. R. Soc. Lond. B* 335:23–30.

Perry, S. 1996. Female-female social relationships in wild white-faced capuchin monkeys, *Cebus capucinus*. *Am. J. Primatol.* 40:167–182.

Petersen, M. R., Beecher, M. D., Zoloth, S. R., Moody, D. B., and W. C. Stebbins. 1978. Neural lateralization of species-specific vocalizations by Japanese macaques (*Macaca fuscata*). *Science* 202:324–327.

Phillips, A. T., Wellman, H. M., and E. S. Spelke. 2002. Infants' ability to connect gaze and emotional expression to intentional action. *Cognition* 85: 53–78.

Pierce, J. 1985. A review of attempts to condition operantly alloprimate vocalizations. *Primates* 26:202–213.

Pinker, S. 1994. *The Language Instinct*. William Morrow and Sons, New York.

Pinker, S. 2002. *The Blank Slate: The Modern Denial of Human Nature*. Viking, New York.

Pinker, S., and P. Bloom. 1990. Natural language and natural selection. *Behav. Brain Sci.* 13:713–783.

Poremba, A., Saunders, R. C., Crane, A. M., Cook, M., Sokoloff, L., and M. Mishkin. 2003. Functional mapping of the primate auditory system. *Science* 299:568–571.

Poremba, A., Malloy, M., Saunders, R. C., Carson, R. E., Herskovitch, P., and M. Mishkin. 2004. Species-specific calls evoke asymmetric activity in the monkey's temporal poles. *Nature* 427:448–451.

Povinelli, D. J., and T. J. Eddy. 1996. What chimpanzees know about seeing. *Monogr. Soc. Res. Child Dev.* 61:1–152.

Povinelli, D. J., and J. Vonk. 2004. We don't need a microscope to explore the chimpanzee's mind. *Mind Lang.* 19:1–28.

Povinelli, D. J., Bierschwale, D. T., and C. G. Cech. 1999. Comprehension of seeing as a referential act in young children, but not juvenile chimpanzees. *Brit. J. Dev. Psych.* 17:37–60.

Powell, A. 1983. *To Keep the Ball Rolling: The Memoirs of Anthony Powell*. Penguin Books, London.

Premack, D. 1975. On the origins of language. Pp. 591–605 in M. D. Gazzaniga and C. B. Blakemore, eds., *Handbook of Psychobiology*. Academic Press, New York.

Premack, D. 1976. *Intelligence in Ape and Man*. Lawrence Erlbaum Associates, Hillsdale, NJ.

Premack, D. 1983. The codes of man and beast. *Behav. Brain Sci.* 6:125–167.

Premack, D., and G. Woodruff. 1978. Does the chimpanzee have a theory of mind? *Behav. Brain Sci.* 4:515–526.

Preston, S. D., and F. B. M. de Waal. 2002. Empathy: Its ultimate and proximate bases. *Behav. Brain Sci.* 25:1–72.

Quine, W. V. O. 1960. *Word and Object*. MIT Press, Cambridge, MA.

Reader, S. M. 2003. Relative brain size and the distribution of innovation and social learning across the nonhuman primates. Pp. 56–93 in D. M. Fragaszy and S. Perry, eds., *The Biology of Traditions: Models and Evidence*. Cambridge University Press, Cambridge.

Reader, S. M., and K. N. Laland. 2002. Social intelligence, innovation, and enhanced brain size in primates. *Proc. Nat. Acad. Sci. USA* 99:4436–4441.

Redford, D. B. 2002. *The Ancient Gods Speak: A Guide to Egyptian Religion*. Oxford University Press, Oxford.

Rendall, D. 2003. Acoustic correlates of caller identity and affect intensity in the vowel-like grunt vocalizations of baboons. *J. Acoust. Soc. Am.* 113:3390–3402.

Rendall, D., Rodman, P. S., and R. E. Emond. 1996. Vocal recognition of individuals and kin in free-ranging rhesus monkeys. *Anim. Behav.* 51:1007–1015.

Rendall, D., Seyfarth, R. M., Cheney, D. L., and M. J. Owren. 1999. The meaning and function of grunt variants in baboons. *Anim. Behav.* 57:583–592.

Rendall, D., Cheney, D. L., and R. M. Seyfarth. 2000. Proximate factors mediating "contact" calls in adult female baboons and their infants. *J. Comp. Psych.* 114:36–46.

Repacholi, B., and A. Gopnik. 1997. Early reasoning about desires: Evidence from 14- and 18-month-olds. *Dev. Psych.* 33:12–21.

Rescorla, R. A. 1988. Pavlovian conditioning: It's not what you think it is. *Am. Psychol.* 43:151–160.

Richards, P. 1995. Local understandings of primates and evolution: Some Mende beliefs concerning chimpanzees. Pp. 265–274 in R. Corbey and B. Theunissen, eds., *Ape, Man, Apeman: Changing Views since 1600*. Leiden University Press, Leiden.

Richards, R. J. 1987. *Darwin and the Emergence of Evolutionary Theories of Mind and Behavior*. University of Chicago Press, Chicago.

Rilling, J. K. 2006. Human and nonhuman primate brains: Are they allometrically scaled versions of the same design? *Evol. Anthr.* 15:65–77.

Rilling, J. K., and T. R. Insel. 1999. The primate neocortex in comparative perspective using magnetic resonance imaging. *J. Hum. Evol.* 37:191–223.

Ritvo, H. 1987. *The Animal Estate*. Harvard University Press, Cambridge, MA.

Rizley, R. C., and R. A. Rescorla. 1972. Associations in second-order conditioning and sensory preconditioning. *J. Comp. Physiol. Psych.* 81:1–11.

Rizzolatti, G., and G. Buccino. 2005. The mirror neuron system and its role in imitation and language. Pp.213–234 in S. Dehaene, J.-R. Duhamel, M. D. Hauser, and G. Rizzolatti, eds., *From Monkey Brain to Human Brain.* MIT Press, Cambridge, MA.

Rizzolatti, G., and L. Craighero. 2004. The mirror-neuron system. *Ann. Rev. Neurosci.* 27:169–192.

Rogers, L., and R. Andrews, eds. 2002. *Comparative Vertebrate Laterality.* Cambridge University Press, Cambridge.

Romanes, G. J. 1881/1977. *Animal Intelligence.* University Publications of America, Washington, D.C.

Rotenberg, K. J. 1982. Development of character constancy of self and other. *Child Dev.* 53:505–515.

Samuels, A., Silk, J. B., and J. Altmann. 1987. Continuity and change in dominance relations among female baboons. *Anim. Behav.* 35:785–793.

Santos, L. R., Flombaum, J. I., and W. Phillips. 2006a. The evolution of human mindreading: How non-human primates can inform social cognitive neuroscience. In S. M. Platek, J. P. Keenan, and T. K. Shackelford, eds., *Evolutionary Cognitive Neuroscience.* MIT Press, Cambridge, MA.

Santos, L. R., Nissen, A. G., and J. A. Ferrugia. 2006b. Rhesus monkeys (*Macaca mulatta*) know what others can and cannot hear. *Anim. Behav.* 71:1175–1181.

Sapolsky, R. M. 1992. Cortisol concentrations and the social significance of rank instability among wild baboons. *Psychoneuroendocrinology* 17:701–709.

Sapolsky, R. M. 1993. The physiology of dominance in stable versus unstable social hierarchies. Pp. 171–204 in W. A. Mason and S. P. Mendoza, eds., *Primate Social Conflict.* State University of New York Press, Albany.

Sapolsky, R. M. 2002. Endocrinology of the stress response. Pp. 409–450 in J. B. Becker, S. M. Breedlove, D. Crews, and M. M. McCarthy, eds., *Behavioral Endocrinology.* MIT Press, Cambridge, MA.

Sapolsky, R. M. 2004. Social status and health in humans and other animals. *Ann. Rev. Anthr.* 33:393–418.

Sapolsky, R. M. 2005. The influence of social hierarchy on primate health. *Science* 308:648–652.

Sapolsky, R. M., Alberts, S. C., and J. Altmann. 1997. Hypercortisolism associated with social subordinance or social isolation among wild baboons. *Arch. Gen. Psych.* 54:1137–1142.

Sapolsky, R. M., Romero, L. M., and A. U. Munck. 2000. How do glucocorticoids influence stress responses?: Integrating permissive, suppressive, stimulatory, and preparative actions. *Endocr. Rev.* 21:55–89.

Sato, N., and K. Nakamura. 2001. Detection of directed gaze in monkeys (*Macaca mulatta*). *J. Comp. Psych.* 115:115–121.

Savage-Rumbaugh, E. S. 1986. *Ape Language: From Conditioned Response to Symbol*. Columbia University Press, New York.

Saxe, R., and L. J. Powell. 2006. It's the thought that counts: specific brain regions for one component of theory of mind. *Psych. Sci.* 17:692–699.

Saxe, R., Carey, S., and N. Kanwisher. 2004. Understanding other minds: Linking developmental psychology and functional neuroimaging. *Ann. Rev. Psych.* 55:87–124.

Scerif, G., Gomez, J.-C., and R. W. Byrne. 2004. What do Diana monkeys know about the focus of attention of a conspecific? *Anim. Behav.* 68:1239–1247.

Scherer, K. R. 1989. Vocal correlates of emotion. Pp. 167–195 in H. Wagner and A. Manstead, eds., *Handbook of Psychophysiology: Emotion and Social Behavior*. John Wiley and Sons, New York.

Schino, G. 2001. Grooming, competition and social rank among female primates: A meta-analysis. *Anim. Behav.* 62:265–271.

Schino, G. 2007. Grooming and agonistic support: a meta-analysis of primate reciprocal altruism. *Behav. Ecol.* 18:115–120.

Schino, G., Geminiani, S., Rosati, L., and F. Aureli. 2004. Behavioral and emotional response of Japanese macaque (*Macaca fuscata*) mothers after their offspring receive an aggression. *J. Comp. Psych.* 118:340–346.

Schino, G., Tiddi, B., and E. Polizzi di Sorrentino. 2006. Simultaneous classification by rank and kinship in Japanese macaques. *Anim. Behav.* 71:1069–1074.

Schino, G., Ventura, R., and A. Troisi. 2003. Grooming among female Japanese macaques: Distinguishing between reciprocation and interchange. *Behav. Ecol.* 14: 887–891.

Schino, G., Ventura, R., and A. Troisi. 2005. Grooming and aggression in captive Japanese macaques. *Primates* 46:207–209.

Schön Ybarra, M. 1995. A comparative approach to the nonhuman primate vocal tract: Implications for sound production. Pp. 185–198 in E. Zimmerman, J. D. Newman, and U. Jurgens, eds., *Current Topics in Primate Vocal Communication*. Plenum Press, New York.

Schusterman, R. J., and D. A. Kastak. 1993. A California sea lion (*Zalophus californianus*) is capable of forming equivalence relations. *Psych. Rec.* 43:823–839.

Schusterman, R. J., and D. A. Kastak. 1998. Functional equivalence in a California sea lion: Relevance to animal social and communicative interactions. *Anim. Behav.* 55:1087–1095.

Schusterman, R. J., and K. Krieger. 1986. Artificial language comprehension and size transposition by a California sea lion (*Zalophus californianus*). *J. Comp. Psych.* 100:348–355.

Schusterman, R. J., Gisiner, R. R., Grimm, B. K., and E. B. Hanggi. 1993. Behavior control by exclusion and attempts at establishing semanticity in marine mammals using match-to-sample paradigms. Pp. 249–274 in H. Roitblat, L. M. Herman, and P. E. Nachtigall, eds., *Language and Communication: Comparative Perspectives*. Lawrence Erlbaum Associates, Hillsdale, NJ.

Schusterman, R. J., Reichmuth Kastak, C., and D. Kastak. 2002. The cognitive sea lions: Meaning and memory in the lab and in nature. Pp. 217–228 in M. Bekoff, C. Allen, and G. Burghardt, eds., *The Cognitive Animal: Empirical and Theoretical Perspectives on Animal Cognition*. MIT Press, Cambridge, MA.

Searcy, W. A., and S. Nowicki. 2005. *The Evolution of Animal Communication: Reliability and Deception in Signaling Systems*. Princeton University Press, Princeton.

Segerstrom, S. C., and G. E. Miller. 2004. Psychological stress and the human immune system: A meta-analytic study of 30 years of inquiry. *Psych. Bull.* 130:601–630.

Seidenberg, M. S., and L. A. Pettito. 1979. Signaling behavior in apes: A critical review. *Cognition* 7:177–215.

Seyfarth, R. M. 1977. A model of social grooming among adult female monkeys. *J. Theor. Biol.* 65:671–698.

Seyfarth, R. M. 1980. The distribution of grooming and related behaviours among adult female vervet monkeys. *Anim. Behav.* 28:798–813.

Seyfarth, R. M., and D. L. Cheney. 1984. Grooming, alliances, and reciprocal altruism in vervet monkeys. *Nature* 308: 541–543.

Seyfarth, R. M., and D. L. Cheney. 1990. The assessment by vervet monkeys of their own and another species' alarm calls. *Anim. Behav.* 40:754–764.

Seyfarth, R. M., and D. L. Cheney. 1997a. Behavioral mechanisms underlying vocal communication in nonhuman primates. *Anim. Learn. Behav.* 25: 249–267.

Seyfarth, R. M., and D. L. Cheney. 1997b. Some general features of vocal development in non-human primates. Pp. 249–273 in M. Husberger and C. T. Snowdon, eds., *Social Influences on Vocal Development*. Cambridge University Press, Cambridge.

Seyfarth, R. M., and Cheney, D. L. 2003. Signalers and receivers in animal communication. *Ann. Rev. Psych.* 54:145–173.

Seyfarth, R. M., Cheney, D. L., and P. Marler. 1980. Vervet monkey alarm calls: Semantic communication in a free-ranging primate. *Anim. Behav.* 28:1070–1094.

Shettleworth, S. 1998. *Cognition, Evolution, and Behavior*. Oxford University Press, Oxford.

Shields, W. E., Smith, J. D., and D. A. Washburn. 1997. Uncertain responses by humans and rhesus monkeys (*Macaca mulatta*) in a psychophysical same-different task. *J. Exp. Psych. Gen.* 126:147–164.

Shields, W. E., Smith, J. D., Guttmannova, K., and D. A. Washburn. 2005. Confidence judgments by humans and rhesus monkeys. *J. Gen. Psych.* 132: 165–186.

Shively, C. A, Register, T. C., Friedman, D. P, Morgan, T. M., Thompson, J., and T. Lanier. 2005. Social stress-associated depression in adult female cynomolgus monkeys (*Macaca fascicularis*). *Biol. Psych.* 69:67–84.

Silk, J. B. 1982. Altruism among adult female bonnet macaques: Explanation and analysis of patterns of grooming and coalition formation. *Behaviour* 79:162–187.

Silk, J. B. 1993. Does participation in coalitions influence dominance relationships among male bonnet macaques? *Behaviour* 126:171–189.

Silk, J. B. 1999. Male bonnet macaques use information about third-party rank relationships to recruit allies. *Anim. Behav.* 58:45–51.

Silk, J. B. 2002. Kin selection in primate groups. *Int. J. Primatol.* 23:849–875.

Silk, J. B. In press a. The adaptive value of sociality in mammalian groups. *Phil. Trans. Roy. Soc. Lond.*

Silk, J. B. In press b. Empathy, sympathy, and prosocial preferences in primates. In R. I. M. Dunbar and L. Barrett, eds., *Oxford Handbook of Evolutionary Psychology*. Oxford University Press, Oxford.

Silk, J. B., Cheney, D. L., and R. M. Seyfarth. 1996. The form and function of postconflict interactions between female baboons. *Anim. Behav.* 52:259–268.

Silk, J. B., Seyfarth, R. M., and D. L. Cheney. 1999. The structure of social relationships among female savannah baboons in Moremi Reserve, Botswana. *Behaviour* 136:679–703.

Silk, J. B., Kaldor, E., and R. Boyd. 2000. Cheap talk when interests conflict. *Anim. Behav.* 59:423–432.

Silk, J. B., Alberts, S. C., and J. Altmann. 2003. Social bonds of female baboons enhance infant survival. *Science* 302:1231–1234.

Silk, J. B., Alberts, S. C., and J. Altmann. 2004. Patterns of coalition formation by adult female baboons in Amboseli, Kenya. *Anim. Behav.* 67:573–582.

Silk, J. B., Brosnan, S. F., Vonk, J., Henrich, J., Povinelli, D. J., Richardson, A. F., Lambeth, S. P., Mascaro, J., and S. J. Schapiro. 2005. Chimpanzees are indifferent to the welfare of other group members. *Nature* 435:1357–1359.

Silk, J. B., Altmann, J., and S. C. Alberts. 2006a. Social relationships among adult female baboons (*Papio cynocephalus*). I. Variation in the strength of social bonds. *Behav. Ecol. Sociobiol.* 61:183–195.

Silk, J. B., Altmann, J., and S. C. Alberts. 2006b. Social relationships among adult female baboons (*Papio cynocephalus*). II: Variation in the quality and stability of social bonds. *Behav. Ecol. Sociobiol.* 61:197–204.

Simon, H. 1974. How big is a chunk? *Science* 183:482–488.

Skinner, B. F. 1956. A case history in scientific method. *Am. Psych.* 11:221–233.

Skinner, B. F. 1974. *About Behaviorism*. Knopf, New York.

Smale, L., Frank, L. G., and K. E. Holekamp. 1993. Ontogeny of dominance in free-living spotted hyaenas: Juvenile rank relations with adult females and immigrant males. *Anim. Behav.* 46:467–477.

Smith, J., Alberts, S. C., and J. Altmann. 2003. Wild female baboons bias their social behaviour towards paternal half-sisters. *Proc. Roy. Soc. Lond. B* 270:503–510.

Smith, J. D. 2005. Studies of uncertainty monitoring and metacognition in animals and humans. Pp. 242–271 in H. S. Terrace and J. Metcalfe, eds., *The*

Missing Link in Cognition: Origins of the Self-Reflective Consciousness. Oxford University Press, New York.

Smith, J. D., Schull, J., Strote, J. and K. McGee. 1995. The uncertain response in the bottlenosed dolphin (*Tursiops truncatus*). *J. Exp. Psych. Gen.* 124: 391–408.

Smith, J. D., Shields, W. E., and D. A. Washburn. 2003. The comparative psychology of uncertainty monitoring and metacognition. *Behav. Brain Sci.* 26: 317–373.

Smith, W. S. 1946. *History of Egyptian Sculpture and Painting.* Oxford University Press, Oxford.

Snowdon, C. T. 1990. Language capacities of nonhuman primates. *Yearbk. Phys. Anthr.* 33:215–243.

Son, L. K., and N. Kornell. 2005. Metaconfidence judgments in rhesus macaques: Explicit versus implicit mechanisms. Pp. 296–320 in H. S. Terrace and J. Metcalfe, eds., *The Missing Link in Cognition: Origins of the Self-Reflective Consciousness.* Oxford University Press, New York.

Sorabji, R. 1993. *Animal Minds and Human Morals.* Cornell University Press, Ithaca, NY.

Spencer, F. 1995. Pithekos to Pithecanthropus: An abbreviated review of changing scientific views on the relationship of the anthropoid apes to humans. Pp. 13–28 in R. Corbey and B. Theunissen, eds., *Ape, Man, Apeman: Changing Views since 1600.* Leiden University Press, Leiden.

Stalnaker, R. 1999. Propositional attitudes. Pp. 678–679 in R. A. Wilson and F. C. Keil, eds., *The MIT Encyclopaedia of the Cognitive Sciences.* MIT Press, Cambridge, MA.

Stavisky, R. C., Adams, M. R., Watson, S. L., and J. R. Kaplan. 2001. Dominance, cortisol, and behavior in small groups of female cynomolgus monkeys (*Macaca fascicularis*). *Horm. Behav.* 39:232–238.

Steiper, M. E., Young, N. M., and T. Y. Sukarna. 2004. Genomic data support the hominoid slowdown and an Early Oligocene estimate for the hominoid–cercopithecoid divergence. *Proc. Nat. Acad. Sci. USA* 101:17021–17026.

Sterck, E. H. M., Watts, D. P., and C. P. van Schaik. 1997. The evolution of female social relationships in nonhuman primates. *Behav. Ecol. Sociobiol.* 41: 291–309.

Stoerig, P., and A. Cowey. 1995. Visual perception and phenomenal consciousness. *Behav. Brain Res.* 71:147–156.

Strong, P. N., and M. Hedges. 1966. Comparative studies in simple oddity learning. I. Cats, raccoons, monkeys, and chimpanzees. *Psychonom. Sci.* 5:13–14.

Struhsaker, T. T. 1967. Auditory communication among vervet monkeys (*Cercopithecus aethiops*). Pp. 281–324 in S. A. Altmann, ed., *Social Communication among Primates.* University of Chicago Press, Chicago.

Subiaul, F., Cantlon, J. F., Holloway, R. L., and H. S. Terrace. 2004. Cognitive imitation in rhesus macaques. *Science* 305:407–410.

Suddendorf, T., and J. Busby. 2003. Mental time travel in animals? *Trends Cog. Sci.* 7:391–396.

Suzuki, K., and T. Kobayashi. 2000. Numerical competence in rats (*Rattus norvegicus*): Davis and Bradford (1986) extended. *J. Comp. Psych.* 114:73–85.

Swift, J. 1726/1984. *Gulliver's Travels.* Shocken Books, New York.

Taylor, M., Esbensen, B. M., and R. T. Bennett. 1994. Childrens' understanding of knowledge acquisition: The tendency for children to report that they have always known what they have just learned. *Child Dev.* 65:1581–1604.

Taylor, S. E., Cousino Klein, L., Lewis, B. P., Gruenewald, T. L., Gurung, R. A., and J. A. Updegraff. 2000. Biobehavioral responses to stress in females: Tend-and-befriend, not fight-or-flight. *Psych. Rev.* 107:411–429.

Terrace, H. S. 1979. *Nim.* Knopf, New York.

Thierry, B. 2000. Covariation of conflict management patterns across macaque species. Pp. 106–128 in F. Aureli and F. B. M. de Waal, eds., *Natural Conflict Resolution.* University of California Press, Berkeley.

Thompson, R. K. R. 1995. Natural and relational concepts in animals. Pp. 175–224 in H. Roitblat and J. A. Meyer, eds., *Comparative Approaches to Cognitive Science.* MIT Press, Cambridge, MA.

Thorsteinsson, E. B., and J. E. James. 1999. A meta-analysis of the effects of experimental manipulations of social support during laboratory stress. *Psych. Health* 14:869–886.

Tinklepaugh, O. L. 1928. An experimental study of representative factors in monkeys. *J. Comp. Psych.* 8:197–236.

Tolman, E. C. 1932. *Purposive Behavior in Animals and Men.* Century Company, New York.

Tomasello, M. 2003. *Constructing a Language: A Usage-based Theory of Language Acquisition.* Harvard University Press, Cambridge, MA.

Tomasello, M. 2005. Beyond formalities: The case of language acquisition. *Ling. Rev.* 22:183–197.

Tomasello, M., and J. Call. 1997. *Primate Cognition.* Oxford University Press, Oxford.

Tomasello, M., Savage-Rumbaugh, E. S., and A. C. Kruger. 1993. Imitative learning of actions on objects by children, chimpanzees, and enculturated chimpanzees. *Child Dev.* 64:1688–1705.

Tomasello, M., Davis-Dasilva, M., Camak, L., and K. K. Bard. 1997. Observational learning of tool use by young chimpanzees. *Hum. Evol.* 2:175–183.

Tomasello, M., Hare, B., and B. Agnetta. 1999. Chimpanzees follow gaze direction geometrically. *Anim. Behav.* 58:769–777.

Tomasello, M., Hare, B., and T. Fogleman. 2001. The ontogeny of gaze following in chimpanzees, *Pan troglodytes,* and rhesus macaques, *Macaca mulatta. Anim. Behav.* 61:335–343.

Tomasello, M., Carpenter, M., Call, J., Behne, T., and H. Moll. 2005. Understanding and sharing intentions: The origins of cultural cognition. *Behav. Brain Sci.* 28:675–691.

Treichler, F., and D. van Tilburg. 1996. Concurrent conditional discrimination tests of transitive inference by macaque monkeys: List linking. *J. Exp. Psych. Anim. Behav. Proc.* 22:105–117.

Tsao, G. Y., Friewald, W. A., Knutsen, T. A., Mandeville, J. B., and R. B. Tootell. 2003. Faces and objects in macaque cerebral cortex. *Nat. Neurosci.* 6:989–995.

Tsao, G. Y., Friewald, W. A., Tootell, R. B., and M. S. Livingston. 2006. A cortical region consisting entirely of face-selective cells. *Science* 311:670-674.

Tulving, E. 1962. Subjective organization in free recall of "unrelated" words. *Psych. Rev.* 69:344–354.

Tulving, E. 2005. Episodic memory and autonoesis: Uniquely human? Pp. 3–56 in H. S. Terrace and J. Metcalfe, eds., *The Missing Link in Cognition: Origins of the Self-Reflective Consciousness.* Oxford University Press, New York.

Uvnas-Moberg, K. 1997. Physiological and endocrine effects of social contact. Pp. 245–262 in S. Carter, I. Lederhendler, and B. Kirkpatrick, eds., *The Integrative Neurobiology of Affiliation.* MIT Press, Cambridge, MA.

van Schaik, C. P. 1983. Why are diurnal primates living in groups? *Behaviour* 87:120–144.

van Schaik, C. P. 2004. *Among Orangutans: Red Apes and the Rise of Human Culture.* Harvard Belknap Press, Cambridge, MA.

van Schaik, C. P., Ancrenaz, M., Borgen, G., Galdikas, B., Knott, C. D., Singleton, I., Suzuki, A., Utami, S. S., and M. Y. Merrill. 2003. Orangutan cultures and the evolution of material culture. *Science* 299:102–105.

Vehrencamp, S. 2000. Handicap, index, and conventional elements of bird song. Pp. 277–300 in Y. Espmark, T. Amundsen, and G. Rosenqvist, eds., *Animal Signals: Signalling and Signal Design in Animal Communication.* Tapir, Trondheim, Norway.

Ventura, R., Bonaventura, M., Koyama, N. F., Hardie, S., and G. Schino. 2006. Reciprocation and interchange in wild Japanese macaques: Grooming, co-feeding, and agonistic support. *Am. J. Primatol.* 68:1–12.

Vick, S.-J., Bovet, D., and J. R. Anderson. 2001. Gaze discrimination learning in olive baboons (*Papio anubis*). *Anim. Cog.* 4:1–10.

Visalberghi, E., and D. Fragaszy. 2002. "Do monkeys ape?": Ten years after. Pp. 471–499 in K. Dautenhahn and C. Nehaniv, eds., *Imitation in Animals and Artifacts.* MIT Press, Cambridge, MA.

Von Fersen, L., Wynne, C., Delius, J., and J. Staddon. 1991. Transitive inference formation in pigeons. *J. Exp. Psych. Anim. Behav. Proc.* 17:334–341.

Wallis, J. D., Anderson, K. C., and E. K. Miller. 2001. Single neurons in the prefrontal cortex encode abstract rules. *Nature* 411:953–956.

Wallman, J. 1992. *Aping Language.* Cambridge University Press, Cambridge.

Walters, J. R., and R. M. Seyfarth. 1987. Conflict and cooperation. Pp. 306–317 in B. B. Smuts, D. L. Cheney, R. M. Seyfarth, R. W. Wrangham, and T. Struh-saker, eds., *Primate Societies*. University of Chicago Press, Chicago.

Warneken, F., and M. Tomasello. 2006. Altruistic helping in human infants and young chimpanzees. *Science* 311:1301–1303.

Washburn, D. A., Smith, J. D., and W. E. Shields. 2006. Rhesus monkeys (*Macaca mulatta*) immediately generalize the uncertain response. *J. Exp. Psych. Anim. Behav. Proc.* 32:185–189.

Wasserman, E. A., and S. L. Astley. 1994. A behavioral analysis of concepts: Application to pigeons and children. Pp. 73–132 in D. L. Medin, ed., *Psychology of Learning and Motivation, Vol. 31*. Academic Press, New York.

Watts, D., and J. Mitani. 2001. Boundary patrols and intergroup encounters in wild chimpanzees. *Behaviour* 138:299–327.

Wechkin, S., Massserman, J. H., and W. Terris. 1964. Shock to a conspecific as an aversive stimulus. *Psychonom. Sci.* 1:47–48.

Weingrill, T., Gray, D. A., Barrett, L., and S. P. Henzi. 2004. Faecal cortisol in free-ranging female chacma baboons: Relationship to dominance, reproductive state, and environmental factors. *Horm. Behav.* 45:259–269.

Weir, A. A. S., Chappell, J., and A. Kacelnik. 2002. Shaping of hooks in New Caledonian crows. *Science* 297:981.

Weiskrantz, L. 1998. Consciousness and commentaries. *Int. J. Psych.* 33: 227–233.

Weiss, D. J., Ghazanfar, A. A., Miller, C. T., and M. D. Hauser. 2002. Specialized processing of primate facial and vocal expressions: Evidence for cerebral asymmetries. Pp. 1–74 in L. Rogers and R. Andrews, eds., *Cerebral Vertebrate Lateralization*. Cambridge University Press, New York.

Whiten, A. 2002. Imitation of sequential and hierarchical structure in action: Experimental studies with children and chimpanzees. Pp. 191–209 in K. Dautenhahn and C. Nehaniv, eds., *Imitation in Animals and Artifacts*. MIT Press, Cambridge, MA.

Whiten, A., Horner, V., and F. B. M. de Waal. 2005. Conformity to cultural norms of tool use in chimpanzees. *Nature* 437:737–740.

Wich, S. A., and H. de Vries. 2006. Male monkeys remember which group members have given alarm calls. *Proc. Roy. Soc. Lond. B* 273:735–740.

Widdig, A., Nuernberg, P., Krawczak, M., Streich, W. J., and F. B. Bercovitch. 2001. Paternal relatedness and age proximity regulate social relationships among adult female rhesus macaques. *Proc. Nat. Acad. Sci. USA* 98:13769–13773.

Wilkinson, J. 1879. *Manners and Customs of the Ancient Egyptians*. Scribner and Welford, New York.

Wilkinson, K. M., Dube, W. V., and W. J. McIlvane. 1998. Fast mapping and exclusion (emergent matching) in developmental language, behavior analysis, and animal cognition research. *Psych. Rec.* 48:407–422.

Williams, J. M., Pusey, A. E., Carlis, J. V., Farm, B. P., and J. Goodall. 2002. Female competition and male territorial behaviour influence female chimpanzees' ranging patterns. *Anim. Behav.* 63:347–360.

Wimmer, H., and J. Perner. 1983. Beliefs about beliefs: Representation and constraining function of wrong beliefs in young children's understanding of deception. *Cognition* 13:103–128.

Winston, J. S., Strange, B. A., O'Doherty, J., and R. J. Dolan. 2002. Automatic and intentional brain responses during evaluation of trustworthiness of faces. *Nat. Neurosci.* 5:277–283.

Wittgenstein, L. 1953. *Philosophical Investigations.* Macmillan, New York.

Wittig, R., and C. Boesch. 2003. The choice of post-conflict interactions in wild chimpanzees (*Pan troglodytes*). *Behaviour* 140:1527–1559.

Wittig, R. M., Crockford, C., Wikberg, E., Seyfarth, R. M., and D. L. Cheney. 2007a. Kin-mediated reconciliation substitutes for direct reconciliation in female baboons. *Proc. Roy. Soc. Lond. B.*

Wittig, R. M., Crockford, C., Seyfarth, R. M., and D. L. Cheney. 2007b. Vocal alliances in chacma baboons, *Papio hamadryas ursinus. Behav. Ecol. Sociobiol.*

Woodward, A. L. 1998. Infants selectively encode the goal object of an actor's reach. *Cognition* 69:1–34.

Worden, R. 1998. The evolution of language from social intelligence. Pp. 148–168 in J. R. Hurford, M. Studdert-Kennedy, and C. Knight, eds., *Approaches to the Evolution of Language.* Cambridge University Press, Cambridge.

Wrangham, R. W. 1977. Feeding behavior of chimpanzees in Gombe National Park, Tanzania. Pp. 504–538 in T. H. Clutton-Brock, ed., *Primate Ecology: Studies of Feeding and Ranging Behaviour in Lemurs, Monkeys, and Apes.* Academic Press, New York.

Wrangham, R. W. 1980. An ecological model of female-bonded primate groups. *Behaviour* 75:262–300.

Wright, A. A., Santiago, H. C., Sands, S. F., and P. J. Urcuioli. 1984. Pigeon and monkey serial probe recognition: Acquisition, strategies, and serial position effects. Pp. 353–374 in H. L. Roitblat, T. G. Bever, and H. S. Terrace, eds., *Animal Cognition.* Lawrence Erlbaum, Hillsdale, NJ.

York, A. D., and T. E. Rowell. 1988. Reconciliation following aggression in patas monkeys, *Erythrocebus patas. Anim. Behav.* 36:502–509.

Zentall, T. R. 1996. An analysis of imitative learning in animals. Pp. 221–244 in C. M. Heyes and B. G. Galef, eds., *Social Learning in Animals: The Roots of Culture.* Academic Press, New York.

Zentall, T. R. 2006. Mental time travel in animals: A challenging question. *Behav. Proc.* 72:173–183.

Zuberbuhler, K. 2000. Referential labeling in Diana monkeys. *Anim. Behav.* 59:917–927.

Zuberbuhler, K. 2001. Predator-specific alarm calls in Campbell's guenons. *Behav. Ecol. Sociobiol.* 50:414–422.

Zuberbuhler, K. 2002. A syntactic rule in forest monkey communication. *Anim. Behav.* 63:293–299.

Zuberbuhler, K. 2003. Referential signaling in non-human primates: Cognitive precursors and limitations for the evolution of language. *Adv. Study Behav.* 33: 265–307.

Zuberbuhler, K. 2005. Linguistic prerequisites in the primate lineage. Pp. 262–283 in M. Tallerman, ed., *Language Origins: Perspectives on Evolution*. Oxford University Press, Oxford.

Zuberbuhler, K., Noe, R., and R. M. Seyfarth. 1997. Diana monkey long distance calls: Messages for conspecifics and predators. *Anim. Behav.* 53:589–604.

Zuberbuhler, K., Cheney, D. L., and R. M. Seyfarth. 1999. Conceptual semantics in a nonhuman primate. *J. Comp. Psych.* 113:33–42.

Index

Page numbers in italics refer to figures. A name without a surname is the name of an individual baboon, unless otherwise specified.

of, 183; by chimpanzees, 143, 279; emergence in children, 202; empathy and, 100; evolution of language and, 270; intentionality and, 147–48; lack of animal evidence for, 204; language learning and, 260, 264–65; in linguistic communication, 244; neural basis of, 124, 183, 192, 201–2; self-awareness and, 199–200, 201–2, 203–4; theory of mind and, 147. *See also* inferences about intentions; inferences about motives; teaching; theory of mind

mental states: challenge to behaviorist view of, 7–9; difficulty of studying, 9; intentionality of, 147–48, 280, 281; propositional attitudes as, 250. *See also* beliefs; intentions; knowledge; mind

metacognition, 200, 201. *See also* self-awareness/consciousness/introspection/metacognition

metaphysics: Darwin and, 1, 3, 4, 273, 276; defined, 2. *See also* baboon metaphysics

Mexican jays, 136

Milford, Canon, 146

Mill, John Stuart: on knowledge, 2; on subjection of women, 50

Miller, Earl, 132

mind: Aristotle on animal behavior and, 18–19; baboon environment and, 12; behaviorists and, 4–5; empiricism vs. rationalism about, 2–4; evolution of, 3–4, 7, 9, 273–76, 282–83; human, 279–83; language and, 249; species-specific predispositions of, 7; twentieth-century views of, 4–10. *See also* mental representations; mental states; theory of mind; thought

mirror neurons, 126, 127, 158

mirror self-recognition: in children, 202; in nonhuman primates, 205

mobbing: of hyena, 47; of leopards, 46–47

Mohembo, 36

Mokopi, Alec, 44

Molomo, 44

mongoose: dwarf, 69; reproductive skew in, 75; suricate, 220, 221

monkeys: artificiality in studies of, 25; brain size in, 123–24; brain specializations in, 124, 126–27; empathy in, 99–100, 193–97; historical views of intelligence in, 120; infanticide in,

57; juvenile dependency period in, 124; mirror tests with, 205; monitoring own knowledge, 206–11, *208, 210;* recognition of dominance ranks in, 91–92; redirected aggression in, 97–98; respect for ownership in, 181, 195; social knowledge of, 144–45; social life as central for, 11–12, 24–25. *See also* baboons; Campbell's monkeys; capuchin monkeys; Diana monkeys; langur monkeys; macaques; marmosets; primates; vervet monkeys

Moremi Game Reserve, 12

Morgan, 209

morphemes, 254

motives: self-awareness and, 201. *See also* inferences about motives; intentions

move grunts, 223, 224, 226; as arbitrary symbol, 261; as discrete signals, 268; emotional arousal and, 229, 232; information content of, 229, *230;* perception of, 231; rules associated with, 268

Mulcahy, N. J., 212

mummification: of baboons by ancient Egyptians, 17; of infant baboon corpse, 195

My Friends the Baboons (Marais), 29, 35

Namaqua people, 31, 34

Namibia: baboon alarm calls in, 165–66; baboons dislodging stones in, 187; goat-herding baboons in, 31, *32,* 33–34; Okavango River in, 36, 37

Nanook, 46

Nat, 104

natural selection: of brain size increase, 15, 124–26; of brain structures and associated behavior, 122–23; Darwin's theory of, 3, 10–11; of intelligence, 121–22; for keeping track of rank and relationships, 117–19; kin selection as, 63–64; language evolution and, 252–53, 270, 271, 280; of large group size in monkeys, 125; mental representations and, 251, 252–53, 270; of mind of baboon, 273–76; of social skills, 110, 125 (*see also* social intelligence hypothesis); of speech capability, 282; of technological intelligence, 141–43. *See also* evolution; reproductive success

Naxos, *63*